WEAVING A NATIONAL MAP

Review of the U.S. Geological Survey Concept of
The National Map

Committee to Review the U.S. Geological Survey Concept of
The National Map

Mapping Science Committee
Board on Earth Sciences and Resources
Division on Earth and Life Studies

NATIONAL RESEARCH COUNCIL
OF THE NATIONAL ACADEMIES

THE NATIONAL ACADEMIES PRESS
Washington, D.C.
www.nap.edu

THE NATIONAL ACADEMIES PRESS, 500 Fifth Street, N.W., Washington, DC 20001

NOTICE: The project that is the subject of this report was approved by the Governing Board of the National Research Council, whose members are drawn from the councils of the National Academy of Sciences, the National Academy of Engineering, and the Institute of Medicine. The members of the committee responsible for the report were chosen for their special competences and with regard for appropriate balance.

This study was sponsored by Award No. 02HQGR0114 between the National Academy of Sciences and the U.S. Geological Survey. Any opinions, findings, conclusions, or recommendations expressed in this publication are those of the authors and do not necessarily reflect the view of the organizations or agencies that provided support for this project.

International Standard Book Number: 0-309-08747-3
Library of Congress Control Number: 2003104244

Copies of this report are available from:

The National Academies Press
500 Fifth Street, N.W.
Washington, DC 20001
Lockbox 285
800-624-6242
202-334-3313 (in the Washington metropolitan area)
http://www.nap.edu

Cover: Layers of geographic information represented as a blanket (upper layer) and quilt (lower layer) covering the conterminous United States. The study committee uses the blanket and quilt metaphor to distinguish components of the USGS *National Map* concept. SOURCE: Susanna Baumgart, University of California, Santa Barbara.

Cover designed by Van Nguyen

THE NATIONAL ACADEMIES
Advisers to the Nation on Science, Engineering, and Medicine

The **National Academy of Sciences** is a private, nonprofit, self-perpetuating society of distinguished scholars engaged in scientific and engineering research, dedicated to the furtherance of science and technology and to their use for the general welfare. Upon the authority of the charter granted to it by the Congress in 1863, the Academy has a mandate that requires it to advise the federal government on scientific and technical matters. Dr. Bruce M. Alberts is president of the National Academy of Sciences.

The **National Academy of Engineering** was established in 1964, under the charter of the National Academy of Sciences, as a parallel organization of outstanding engineers. It is autonomous in its administration and in the selection of its members, sharing with the National Academy of Sciences the responsibility for advising the federal government. The National Academy of Engineering also sponsors engineering programs aimed at meeting national needs, encourages education and research, and recognizes the superior achievements of engineers. Dr. Wm. A. Wulf is president of the National Academy of Engineering.

The **Institute of Medicine** was established in 1970 by the National Academy of Sciences to secure the services of eminent members of appropriate professions in the examination of policy matters pertaining to the health of the public. The Institute acts under the responsibility given to the National Academy of Sciences by its congressional charter to be an adviser to the federal government and, upon its own initiative, to identify issues of medical care, research, and education. Dr. Harvey Fineberg is president of the Institute of Medicine.

The **National Research Council** was organized by the National Academy of Sciences in 1916 to associate the broad community of science and technology with the Academy's purposes of furthering knowledge and advising the federal government. Functioning in accordance with general policies determined by the Academy, the Council has become the principal operating agency of both the National Academy of Sciences and the National Academy of Engineering in providing services to the government, the public, and the scientific and engineering communities. The Council is administered jointly by both Academies and the Institute of Medicine. Dr. Bruce M. Alberts and Dr. Wm. A. Wulf are chairman and vice chairman, respectively, of the National Research Council.

www.national-academies.org

COMMITTEE TO REVIEW THE U.S.GEOLOGICAL SURVEY CONCEPT OF THE *NATIONAL MAP*

KEITH C. CLARKE, *Chair,* University of California, Santa Barbara
MICHAEL R. ARMSTRONG, City of Des Moines, Iowa
DAVID J. COWEN, University of South Carolina, Columbia
DONNA P. KOEPP, Harvard University, Cambridge, Massachusetts
XAVIER LOPEZ, Oracle Corporation, Nashua, New Hampshire
RICHARD D. MILLER, Kansas Information Technology Office, Topeka
GALE W. TESELLE, USDA/NRCS (*retired*), Mitchellville, Maryland
WALDO R. TOBLER, University of California, Santa Barbara
NANCY VON MEYER, Fairview Industries, Pendleton, South Carolina

National Research Council Staff

PAUL M. CUTLER, Study Director
EILEEN M. MCTAGUE, Research Assistant
RADHIKA S. CHARI, Senior Project Assistant
MOHAN SEETHARAM, Intern (*Until August, 2002*)

TERESIA K. WILMORE, Project Assistant
WINFIELD SWANSON, Editor

Acknowledgments

This report has been reviewed in draft form by individuals chosen for their diverse perspectives and technical expertise, in accordance with procedures approved by the National Research Council's Report Review Committee. The purpose of this independent review is to provide candid and critical comments that will assist the institution in making its published report as sound as possible and to ensure that the report meets institutional standards for objectivity, evidence, and responsiveness to the study charge. The review comments and draft manuscript remain confidential to protect the integrity of the deliberative process. We wish to thank the following individuals for their review of this report:

Chaitan Baru, University of California, San Diego
Donald Cooke, Geographic Data Technology, Lebanon, New Hampshire
Michael Furlough, University of Virginia, Charlottesville
Mary Gunnels, U.S. Department of Transportation, Washington, D.C.
Alan Leidner, New York City Department of Information Technology and Telecommunications, New York
Mark Schaefer, NatureServe, Arlington, Virginia
Nancy Tosta, Ross & Associates Environmental Consulting, Seattle, Washington.
Eugene Trobia, Arizona State Cartographer, Phoenix, Arizona

Although the reviewers listed above have provided many constructive comments and suggestions, they were not asked to endorse the conclusions or recommendations nor did they see the final draft of the report before its release. The review of this report was overseen by Debra Knopman, RAND. Appointed by the National Research Council, she was responsible for making certain that an independent examination of this report was carried out in accordance with institutional procedures and that all review comments were carefully considered. Responsibility for the final content of this report rests entirely with the authoring committee and the institution.

Contents

Executive Summary

This report responds to a request by the U.S. Geological Survey (USGS) that the National Research Council (NRC) review its concept of *The National Map*. *The National Map* is envisioned by USGS as a database providing "public domain core geographic data about the United States and its territories that other agencies can extend, enhance, and reference as they concentrate on maintaining other data that are unique to their needs" (USGS, 2001).

A motivation behind *The National Map* is the need to update aging national paper map coverage. Since 1807 the United States has recognized a federal government responsibility to develop and disseminate maps and charts to "promote the safety and welfare of the people" (Thompson, 1988). From its inception in 1879 the USGS has developed a central role in mapping the nation. Today the USGS's primary topographic map series includes more than 55,000 unique map sheets, and 220,000 digital images. Although this ranks as a great, if unsung, scientific accomplishment, most of the nation's map coverage is out of date. Paper map sheets in USGS's primary map series are, on average, 23 years old (USGS, 2001). Map timeliness, a function of the rapidity and effectiveness of map revision, remains a critical national need. At the same time, events such as the September 11, 2001, terrorist attacks and recent natural disasters have shown that current information can save lives, and protect public and private property. The demand is great for up-to-date information in real time for public welfare and safety.

In July 2002 the NRC appointed the Committee to Review the U. S. Geological Survey Concept of *The National Map* (see Appendix A). This committee acted under the auspices of the Mapping Science Committee. The committee's charge was to review the goals for *The National Map* and evaluate the approaches described in existing USGS documents to meet these goals, the potential benefits of *The National Map* to the nation (e.g., for homeland security) and the role of the USGS as the proposed leader of this effort. Specific aspects to be evaluated were (1) the proposed data characteristics and recommended methods for providing consistent data for these characteristics over areas of arbitrary geographic size or shape from multiple data holdings whose characteristics will vary among sources; (2) the means described in existing USGS documents to encourage widespread use of *The National Map* through low-cost data in the public domain, and still encourage participation in data maintenance by public, private, and nonprofit organizations; and (3) the roles described for the USGS and partners, including volunteers, to undertake the project.

Information gathering for the study was centered on a workshop that took place in Washington, D.C., on September 25-26, 2002. Approximately 40 people from all levels of government, the private sector, nongovernmental organizations, and other stakeholders were invited to submit written input prior to their attendance at the workshop (see Appendix B). These submissions are accessible at <http://www7. nationalacademies.org/besr/National_Map_Participants.html>, and some key points are extracted as quotes in Appendix D. In addition to the written input and discussions at the workshop, the committee drew on materials provided by the USGS, including its "vision" document for *The National Map* (USGS, 2001) and a compilation of community input solicited by USGS on the previous draft of its vision document.

BROAD CONCLUSIONS AND KEY RECOMMENDATIONS

The committee acknowledges that the USGS Geography Discipline[1] has made a bona fide effort to confront its future head-on with *The National Map* vision. If successful, the program will have great benefits to the nation. The *National Map* vision of the USGS is ambitious, challenging, and worthwhile. Nevertheless, there is also a uniform sense that the project is not well defined and needs a thorough definition.

---

[1]The USGS has recently changed the nomenclature of its branches to emphasize its scientific disciplinary structure. The National Mapping Division and Geography Research are now reflected at USGS by the Geography Discipline.

Technically the project may be feasible; organizationally it will require a significant investment in restructuring and rethinking the systems that have changed little over the last two decades. The USGS vision document adds to the mix of already complex programs and terminologies, and reads as a USGS-specific document rather than the concept document for a compelling new national program that reaches far beyond a single federal agency. There is little new in the vision document that has not already been written or discussed as part of the National Spatial Data Infrastructure (NSDI) and the Framework program, and beyond what is already mandated for USGS by the recently revised Office of Management and Budget (OMB) Circular No. A-16. Furthermore, there is at least a 10-year history of recommendations (see Appendix C) directed mostly at the USGS to build partnerships and Framework data,[2] and yet the mapping mission is still at the conceptual stage. Some of the earlier ideas, if implemented before today, could have led to completion of *The National Map*, as outlined.

Recommendation: The USGS should move expeditiously to develop an implementation strategy for its *National Map* concept in collaboration with USGS's many partners. The strategy should be clear on the needs, roles, incentives, and projected costs for all partners, on goals, milestones, and responsibilities, and on the USGS role with respect to Federal Geographic Data Committee (FGDC) activities, Geospatial One-Stop, and other initiatives to build out the National Spatial Data Infrastructure. The draft implementation plan should be circulated to all FGDC members and partners for comment.

The National Map as presently conceived is a large, ambitious project. Its success depends upon a number of factors that are beyond the control of the USGS. As a general approach to the project, the USGS should continue to build from a more modest, step-wise series of activities that are readily attainable, such as its pilot projects. The committee sees the development of integrated base geographic information for the nation as a cultural and institutional challenge more than a scientific or technical one. Tackling this challenge will require (1) the USGS Geography Discipline to be proactive in developing relationships at all levels of government, (2) significant engagement by USGS leadership, and (3) that the USGS critically examine its philosophy, structure, and processes. This new role is distinct from and builds upon the USGS's existing

---

[2]Cadastral data, digital orthoimagery, elevation and bathymetry, geodetic control, government units, hydrography, and transportation.

coordination role. The coordination role remains necessary, particularly in the areas of standards development and quality assurance, but a key question the USGS must ask its partners at every government level is how can the USGS assist them, and are these partners willing to provide resources to support the resulting identified needs and demands?

Recommendation: The USGS should make a priority of building the necessary partnerships for an integrated spatial database, while continuing to use small steps and pilot studies to gain experience in revision, integration, and updating procedures and partnerships. The pilot studies should be seen not only as technical but also as organizational and management prototypes. The USGS should place more Geography Discipline emphasis on building these partnerships to assemble Framework data through collaborative programs.

In this report the committee discusses the similarities of *The National Map* and the existing National Atlas of the United States of America®.[3] The latter is described by the USGS as being a component of *The National Map* (USGS, 2001). In large part the National Atlas has been built using coordination and partnerships, using a national standard to develop nationally consistent small-scale databases from larger-scale data. Data themes are owned and maintained by different federal agencies and updates are provided to the USGS for inclusion in the National Atlas. The same should be true of *The National Map*, though at larger scales and with more partners.

The USGS concept of *The National Map* has two principal components, each dependent on the other. The first is a nationally consistent digital map coverage maintained at one or more uniform scales. The second is a patchwork of varied scales including high-resolution local data. We use a blanket and patchwork quilt metaphor in explaining these two components (see Figure 2.1). The blanket, which we term the enhanced National Atlas (to extend the existing program), would be built with public domain data and broadly disseminated following the philosophy in OMB Circular No. A-16. The second component, the patchwork quilt *National Map*, would be the result of contributed imagery and maps from local, state, and tribal governments, and from private and nonprofit organizations, contributed as part of a sweeping collaborative effort. This quilt would consist of patches of larger-scale data adhering to national standards but with varied resolutions and filled with smaller-scale data from the enhanced National Atlas when no other source exists.

---

[3]The committee refers to this as the "National Atlas" throughout the report.

Some of the data will be public, some proprietary with publicly accessible metadata.

The USGS would serve as the integrator for all map contributions, assembling and merging data, and certifying and issuing a "seal of approval" to data included in *The National Map* or as an update in the nationally consistent enhanced National Atlas. The USGS goal of seven-day updates could be attainable using this schema. Such a dynamic *National Map* will need to support multiple scales, resolutions, classifications, and feature types provided by *National Map* partners. It will also require extraordinary coordination.

Recommendation: Two synergistic organizational structures are needed for the USGS's contribution to building the National Spatial Data Infrastructure. The first is an enhancement of the existing National Atlas and includes Framework data (some of which already exists and will require partnerships with the National Oceanic and Atmospheric Administration [NOAA] and the Bureau of Land Management [BLM] in particular). The data in the atlas should be public domain, at such a consistent scale as 1:12,000 or 1:24,000, and could be served through many existing and new gateway public and private Internet sites. The second structure, called *The National Map*, would serve users needing integrated larger-scale data, drive updates to the enhanced National Atlas, and implement many of the ideas that the USGS has proposed: seamlessness, voluntary contribution, a mix of public domain and private data, shared metadata, and nonuniform scale.

In addition to the key recommendations included in this summary the committee adds further recommendations in Chapter 5. The recommended actions could, when embraced and implemented, ensure that the USGS enters the twenty-first century with a sound national mapping strategy.

1

Introduction

HISTORY

Since 1807 the United States has recognized a federal government responsibility to develop and disseminate maps and charts to "promote the safety and welfare of the people" (Thompson, 1988). This mission was significantly expanded in 1879, when Congressman Hewitt of New York in authoring legislation to create the United States Geological Survey (USGS) asked, "What is there in this richly endowed land of ours which may be dug, or gathered, or harvested, and made part of the wealth of America and of the world, and how and where does it lie?" Charged with this task, Clarence King as the first director of the USGS concluded that it mandated the mapping of the nation, a task increasingly formalized as a defining operation for the USGS. As the second director, John Wesley Powell, acting on advice from the National Academy of Sciences (NAS, 1884), consolidated national mapping efforts and started work on a 20-year mapping program as a "sound framework for scientific study and national resource development." This initially consisted of maps at 1:250,000 and 1:125,000 scale, but by 1894 two-thirds of the nation was covered at the more detailed scale of 1:62,500. During the twentieth century the mapping evolved into a National Mapping program, today the responsibility of USGS's Geography Discipline (see Box 1.1).

During the 1950s a standard mapping scale of 1:24,000 was adopted for the contiguous states and Hawaii, and the national mapping effort started

---

BOX 1.1
Mission of the USGS National Mapping Program

"To accomplish its mission, the Office of the Associate Director for Geography conducts the National Mapping Program to meet the Nation's need for basic geospatial data, ensuring access to and advancing the application of these data and other related earth science information for users worldwide. The responsibility of the National Mapping Program is to ensure the production and availability of basic cartographic and geographic spatial data of the country; coordinate national geospatial data policy and standards; provide leadership for the management of earth science data and for information management; acquire, process, archive, manage, and disseminate the land remote sensing data of the Earth; and improve the understanding and application of geospatial data and technology."

SOURCE: Michael Domaratz, USGS, personal communication of a planned revision to the Department of the Interior manual, 2002.

---

anew based on the principles of photogrammetry and air photograph interpretation pioneered and advanced during World War II. By 1991 the 1:24,000-scale paper topographic base map of the United States was complete, although Alaska remained mostly covered at the 1:63,360 scale.

Today the USGS's primary topographic map series includes more than 55,000 unique map sheets and 220,000 digital orthorectified images. Although this ranks as a great, if unsung, scientific accomplishment, most of the nation's map coverage[1] is out of date. Paper map sheets in the USGS's primary map series are on average 23 years old (USGS, 2001). Map timeliness, a function of the rapidity and effectiveness of map revision, remains a critical national need. At the same time, events such as the September 11, 2001, terrorist attacks and recent natural disasters have shown that current information in the public domain can save lives and protect public and private property. The demand is great for up-to-date information for public welfare and safety.

Yet even before the completion of the monumental undertaking of mapping the nation, mapping methodology began to transform itself around the digital information revolution.[2] Paper maps now meet only a

---

[1]Geographic extent of coordinates in the map database.
[2]Hence, the committee uses the term map, as in *The National Map*, not in the strict sense of a paper graphic depiction of a scaled symbolization of geographical

minority of the needs of map users, and maps have evolved into digital spatial data, computer representations of the nation's features and of the paper maps that first captured them. The digital transformation has taken the mapping world by storm. The USGS began a digital conversion effort in the 1970s, and digital mapping of America is taking place at the local level as literally thousands of separate private, nonprofit, academic, tribal, state, and local organizations build and now maintain their own detailed maps and databases to suit their own needs, often independent of federal mapping efforts.

Recognizing this transition, the USGS envisions its role as the national spatial data coordinator, rather than as the central map creator for the next level of the mapping of the nation (USGS, 2001). Key to this role will be the USGS's proposed leadership in spatial data standards development (USGS, 2001).[3] The broader aim is to build out the National Spatial Data Infrastructure (NSDI) (see Box 1.2) that links spatial data holdings across the nation. In the committee's view this infrastructure will approach completion once it contains high-quality, integrated, nationally consistent data linked to other larger-scale spatial datasets, such as those being generated by communities around the country. Hundreds of millions of dollars worth of data and technology reside with local, state, and tribal entities, and thousands of data-collection projects are planned. What is needed is the collaborative architecture and incentives to maximize the benefit to all participants.

CURRENT CONTEXT

The USGS has proposed a new national map.[4] In the vision document proposing the project (USGS, 2001) *The National Map* is seen as provid-ing "public domain core geographic data about the United States and its territories that other agencies can extend, enhance, and reference as they

---

features in their correct locations, but as any such depiction, and even a "latent" map within a spatial database. Thus a "national" map could exist independently of any scheme for the delivery of hard copy scaled depictions of part of the map coverage.

[3]This role is already assigned to the USGS in OMB Circular No. A-16. As a lead agency on a number of data themes the USGS has the responsibility to facilitate the development and implementation of Federal Geographic Data Committee (FGDC) standards in those themes.

[4]The generic term "national map" (which in the modern context of maps is centered on a digital database) is distinguished from the USGS program for *The National Map* by capitalizing and italicizing the latter.

10

WEAVING A NATIONAL MAP

BOX 1.2
The National Spatial Data Infrastructure

Called for in Executive Order 12906 (see Appendix C), the goal of the NSDI is to reduce duplication of effort among agencies, improve quality and reduce costs related to geographic information, to make geographic data more accessible to the public, to increase the benefits of using available data, and to establish key partnerships with states, counties, cities, tribal nations, academia, and the private sector to increase data availability.

The NSDI has come to be seen as having the technology, policies, criteria, standards, and people necessary to promote spatial data sharing throughout all levels of government, the private and nonprofit sectors, and academia. It provides a base or structure of practices and relationships among data producers and users that facilitates data sharing and use. It includes a set of actions and new ways of accessing, sharing, and using geographic data that enables comprehensive analysis of data to help decisionmakers chose the best course(s) of action. One focus of NSDI activities has been on Framework data themes, which are cadastral data, digital orthoimagery, elevation and bathymetry, geodetic control, government units, hydrography, and transportation. The Federal Geographic Data Committee (FGDC) is the primary coordinator of the NSDI at the national level.

Much has been accomplished in recent years to further the implementation of the NSDI, but there is still much to be done to achieve the vision of current and accurate geographic data being readily available across the country.

---

concentrate on maintaining other data that are unique to their needs." *The National Map* is seen as a spatial database rather than a traditional map. USGS's new role in *The National Map* is as the integrator of locally held spatial data sets. The USGS would be, in other words, the agency responsible for managing the placement of localized data into a common reference frame.

The USGS has presented *The National Map* as its way to continue to serve the mission of the National Mapping program in the twenty-first century. Indeed, it could be said that the USGS National Mapping program's long-term survival depends on the success of this potentially invaluable resource, which will enhance the NSDI. The USGS initiative is not alone on the federal side in seeking to build the NSDI. The Federal Geographic Data Committee (see Box 1.3) is accelerating the advancement of the NSDI, driven by the proposed Internet portal Geospatial One-Stop (see Box 1.4).

HIGHLIGHTS OF THE USGS VISION DOCUMENT

This section summarizes the characteristics of *The National Map* vision as laid out by the USGS in its vision document (USGS, 2001).

Data Characteristics

National Map layers will include

- high-resolution digital orthoimagery;
- high-resolution elevation and bathymetry;
- vector data themes (hydrography, transportation, structures, boundaries of government features and publicly owned lands);
- geographic names (e.g., physical and cultural features); and
- land-cover information.

National Map data characteristics will include

- up-to-date data (with the ultimate goal that none will be more than seven days older than a change on the landscape);
- seamlessness (accessible for arbitrarily defined study areas);
- consistent classification;
- variable resolution (never worse than existing primary series topographic maps for an area);
- completeness;
- consistency and integration;
- variable positional accuracy;
- reprojection to a uniform datum;
- standardized content;
- standardized metadata; and
- appropriate archiving to capture the temporal dimension of change in data layers.

BOX 1.3
Structure, Membership, and Current Role of the Federal Geographic Data Committee

The Federal Geographic Data Committee (FGDC) is an interagency committee that coordinates the development, use, sharing, and dissemination of geographic data nationally. The OMB established the committee in the 1990 revision of Circular No. A-16 and reestablished it in the circular's 2002 revision. The FGDC evolved in part from an earlier committee established by OMB in 1983 called the Federal Interagency Coordinating Committee on Digital Cartography.

The FGDC membership is composed of representatives from 18 cabinet-level and independent federal agencies. Eleven nonfederal collaborating partners representing state and county governments, tribal governments, cities, academic institutions, geographic information science professionals, and the private sector are active stakeholders in FGDC.

FGDC is organized into a Steering Committee, Coordination Group, 12 data subcommittees, and 13 working groups. The Steering Committee sets strategic direction for the FGDC as a whole and the Coordination Group advises on day-to-day business. As laid out in OMB Circular No. A-16, the Steering Committee is chaired by the secretary of the interior or designee and vice-chaired by the deputy director of management for OMB, or designee. Departments are most often represented by cabinet-level appointees, independent agencies by managers of the geospatial and technology programs, and the nonfederal collaborating partners by the leaders of their organizations. The monthly Coordination Group meeting is chaired by the FGDC staff director. The FGDC subcommittees are organized by data themes and coordinate the development, use, and dissemination of that data theme. Working groups play a crosscutting role, dealing with issues that span many of the subcommittees. Staff support for FGDC committees is provided by the FGDC secretariat staff. FGDC receives a budget of approximately $4.9 million from the USGS to facilitate coordination, develop standards, sponsor cooperative agreements, and support the secretariat staff.

Some of the accomplishments attributed to the FGDC include the development and issuance of approximately 25 spatial data standards including the metadata standard; establishment of clearinghouses that provide access to spatial data using metadata standards; hundreds of cooperative agreements that sponsor the development of Framework data, the development and testing of spatial data access technology and interoperability; and numerous publications and educational materials describing the National Spatial Data Infrastructure and its various components.

SOURCE: FGDC (2002a); Milo Robinson, FGDC, personal communication, 2002.

BOX 1.4
What is Geospatial One-Stop?

Geospatial One-Stop, one of twenty-four e-government initiatives of the Office of Management and Budget, includes a proposed portal that is designed to be the single online portal for the government's spatial information databases (Bhambani, 2002). The One-Stop project aims to accelerate ongoing efforts to build the National Spatial Data Infrastructure and "to spatially enable the delivery of government services (FGDC, 2002b)." The project has a finite lifespan, initially intended to be two years, and is focused on the seven Framework data themes (see Box 1.2).

For Geospatial One-Stop to be effective, participating data producers must classify and document their data holdings following accepted standards. Geospatial One-Stop aims to advance spatial data collaboration initiatives by accomplishing the following tasks (FGDC, 2002c):

1. Develop and implement data and metadata standards for NSDI Framework data themes;
2. Maintain an operational inventory of spatial data and publish their metadata records through the portal;
3. Publish metadata detailing planned data development, acquisition, and update investments;
4. Deploy enhanced data access and Web-mapping services; and
5. Establish a comprehensive electronic portal as a logical extension to the NSDI clearinghouse network.

Geospatial One-Stop is an ambitious project with a challenging schedule. Data standards for Framework data themes are slated for completion in 2003. To manage the initiative, a Geospatial One-Stop Board of Directors has been established that brings together representatives of multiple federal agencies, and state, local, and tribal governments organizations.

SOURCES: Bhambani (2002); Cameron (2002); FGDC (2002b,c).

Widespread Use and Maintenance of Data

The National Map will have a range of traits to encourage widespread use and maintenance. The vision document proposes that *The National Map* will

- initially draw heavily on existing sources of data;
- update databases only with needed changes;
- rely on notification of changes from those closest to the change (e.g., local governments, certified local volunteers);
- comprise distributed networked data sources maintained to common standards by data holders;
- be available free across the Internet (or at a cost of media and distribution);
- use industry-supported, open, standards-based protocols;
- provide unrestricted and immediate access to data;
- maintain data in the public domain;
- be readily linked to databases of other federal agencies and organizations, acting perhaps as a reference base on which to link other attribute information, or combining data displays with those of other organizations; and
- provide access to detailed data or value-added services for a fee from public and private organizations.

Roles of the USGS and Partners

In the vision document USGS roles are

- guarantor of completeness, consistency, and accuracy;
- responsible party for awareness, availability, and utility of the map database;
- catalyst and collaborator for creating and stimulating partnerships;
- integrator and certifier of data from contributors;
- owner and producer of content when no other suitable and verifiable source exists;
- leader in the development and implementation of national spatial data standards (additionally, the USGS will ensure data quality through standards development, by devising and implementing quality assurance procedures, and by promoting process certification criteria for content providers);

• provider of liaison staff to other agencies and in area maintenance offices to be closer to users and partners to understand their needs; and

• provider of field centers to provide management, technical, and administrative support to *National Map* activities.

Specific outcomes assigned in the vision document are

• a federal advisory committee (with membership from all sectors) will advise on evolving requirements, approaches to maintenance and processing, systems and technology development and implementation, and skill enhancements;

• federal partners will collaborate with the USGS in data development and maintenance (e.g., Bureau of the Census for roads and boundaries; Department of Agriculture for imagery; Environmental Protection Agency for hydrography and land cover; Federal Emergency Management Agency for elevation; National Oceanic and Atmospheric Administration for bathymetry and geodetic control; National Imagery and Mapping Agency for data needed for national defense);

• state, tribal, regional, and local partnerships will be strengthened to bring high-resolution data into *The National Map* (cooperative agreements will support activities that are mutually beneficial to state or local partners and the USGS);

• private-sector partnerships will be sought, particularly in supplying analysis and visualization tools, for research on cartographic technologies, and for open technology and processing standards;

• the USGS will acquire data from the private sector with licensing provisions that support broad access and use;

• academic and library partners will collaborate with the USGS on cartographic, geographic, and general information science challenges (the USGS will continue its partnership with the U.S. Government Printing Office on assuring access to *The National Map* through the Federal Depository Library program [FDLP], and will seek library partnerships for data archiving for permanent access); and

• public partnerships will arise in the form of certified volunteers possessing Global Positioning System (GPS) units.

THE CHARGE TO THE COMMITTEE

The National Research Council was asked by the USGS to review its concept of *The National Map* as part of the broader review process of their vision document (USGS, 2001). The three focal topics for the committee parallel the structure of the vision document. Centered on a workshop, the study reviewed the goals for *The National Map* and evaluated approaches described in existing USGS documents to meet these goals, the potential benefits of *The National Map* to the nation (e.g., for homeland security) and the role of the USGS as the proposed leader of this effort. Specific aspects evaluated were (1) the proposed data characteristics and recommended methods for providing consistent data for these characteristics over areas of arbitrary geographic size or shape from multiple data holdings whose characteristics will vary among sources; (2) the means described in USGS documents to encourage widespread use of *The National Map* through low-cost data in the public domain, and still encourage participation in data maintenance by public, private, and nonprofit organizations; and (3) the roles described for the USGS and partners, including volunteers, to undertake the project.

The National Research Council's Committee to Review the U.S. Geological Survey Concept of *The National Map* has taken its task literally, that is, we have reviewed not only the USGS's vision in their descriptive document (USGS, 2001) but also the concept and implications of a national "map" independent of the USGS's document. The USGS vision document has undergone extensive review and public comment, and the committee began its assigned task by reviewing these earlier inputs. All involved with the committee's review and the workshop believe that the USGS has created a vision of profound scope with critical national impacts. As with the 1884 National Academy of Sciences report, the committee seeks a "perfect co-ordination and co-operation" between the parties involved that will bring forth a "practicable ... plan for surveying and mapping the Territories of the United States on such general system as will, in their judgment, secure the best results for the least possible cost."

REPORT STRUCTURE

This report begins with a discourse of the need for a national map, continues with a discussion of it's component parts and how such a concept could be implemented, and concludes with recommendations. The recommendations are grouped into the three focus areas in the committee's charge. Supplemental information has been placed in appendixes, but relevant descriptions of programs, concepts, and terminology necessary to follow the flow of discussion have been included in boxes that parallel the report.

During its deliberations the committee considered the comments received in written and oral form during the workshop. Where pertinent, these comments are quoted and attributed accordingly. These statements have been chosen to illustrate particular points or to reflect the consensus of the committee.

2

The Need for a National Map

INTRODUCTION

This chapter reviews the need for and stakeholders in a national map, the policy context for the USGS role in building such a map, and related successful federal programs from which lessons could be learned. Without a well-defined need there will not be clear incentives for others to be partners with the USGS to build, maintain, and use *The National Map*. Building the partnerships that the USGS proposes is a complex and demanding task.

STAKEHOLDERS AND THEIR NEEDS

From the USGS perspective *The National Map* (USGS, 2001) is needed to meet federal mapping requirements, to reduce unnecessary redundancy in federal mapping efforts, to continue to serve the needs of paper-map users, and to ensure standards for digital and paper maps. From the committee's perspective there is a positive and significant benefit to the nation from coordinated, timely, and accurate topographic mapping. This benefit pertains to the myriad users at all levels, including that of the individual citizen. General facts about the surface of the Earth within the confines of the United States and its territories are public domain knowledge. These facts are of primary use to the assurance of public safety during natural and human-induced disasters; can assure the effective use and protection of the nation's resources; and support literally hundreds of applications that contribute to every citizen's daily life. Accurate and timely map data aid the hunter, the boater, the letter carrier, the schoolchild, the city manager, the law enforcement officer, the scientist, and the President. In this way little has changed since 1807.

The value of *The National Map* approach for homeland security is apparent. Homeland security information must reside locally and be integratable nationally (e.g., see comments of Hugh Bender, Appendix D). When rapid response is needed in the critical first minutes and hours, local information is used by local first responders (e.g., see comments of Scott Oppmann and Ian von Essen, Appendix D), but when looking for patterns, trends, and tendencies, the national view will be critical. This was the case with the terrorist attacks on September 11, 2001, when local data in the form of New York City's NYCMap (Keller and Kreizman, 2002) were prominent in helping in the rescue, recovery, and cleanup at the World Trade Center. As operations continued the federal role was to add new data and integrated information beyond the municipal boundaries of New York City.

All of these needs and applications must be balanced against the reality that *The National Map* will need to have a clear direction and focus and that it cannot be all things to all people and all applications (e.g., see comments of Yves Belzile, Appendix D), because this will generate unreasonable expectations and skepticism from data producers. In particular the committee distinguished between functions of unified coverage at a common scale, where the main contribution is to remove the arbitrary limitations of local districts through common specifications and extent versus the desire at all times to have the most accurate and up-to-date information, regardless of common standards or scale.

What do participants in the planning process for *The National Map* imagine it to be? During the workshop for this study, participants were asked to visualize and describe *The National Map* at some time in the future. The following vignette based on their input illustrates the range of stakeholders and process managers in *The National Map*.

It is now 2007 and the nation awaits preparation for the 2010 decennial census. Updating the TIGER (Topologically Integrated Geographic Encoding and Referencing) files for the 2000 census was a slow, massive, and costly undertaking, but there has now been a watershed change in thinking about how to coordinate the development of a consistent set of street centerlines for digital maps. New streets are routinely submitted in digital format by developers to municipal and county government offices to ensure that they are given unique names and have logical street address numbering. These digital files are inserted into the county's database. Once the county's geodetic coordinator has verified the positional accuracy of the data and inserted a time stamp, this transaction is automatically sent to the state area integrator. At the state level the transaction is integrated with other transportation map themes. The state area integrator's office automatically sends a notice to the state update subscribers who have requested notification of such transactions. The new roads are inserted into the E911 system. The transaction is then automatically forwarded to the USGS National Map coordinating office.

As guarantor of consistency and integration the National Map office runs a conflation routine and error-checking system to insert unique numbers for the street intersections and features according to the FGDC transportation feature standard. When these verification checks are completed, the new streets are inserted into the "online" version of the enhanced National Atlas, accessible immediately through Geospatial One-Stop, the Geography Network, and other distribution services, each of which determines whether it is time to update their distributed versions of the national road map layer. The National Atlas layers are backward integrated into the larger National Map that includes access to some proprietary data sources. The private sector can now redistribute the new roads data for vehicle navigation systems, location-based services, and specialized systems for the disabled community. Libraries and other on-demand map producers check The National Map to ensure that it is as current as possible. The entire data collection, verification, and distribution system is controlled by an approved workflow process and the system is virtual and transparent to the end user. The private sector has also developed a number of plug-ins for Web browsers that enable the users to develop their own user profiles, map design preferences, and specialized applications. There is a tab on the Google website for NSDI access.

Over the past five years several institutional changes have occurred. Geospatial One-Stop accelerated the completion of Framework data themes for the nation. Every thematic layer has accepted standards for content, attributes, spatial resolution, and accepted protocols for data capture. All federal agencies have agreed to use one consistent regional scheme to deal with states. The states have been able to acquire the updated address records from the utilities to ensure that the E911 system is complete, current, and accurate. Most states have developed their own system for aiding local governments, and 30 states have integrated street centerlines and addresses. There is a federal grant program to provide the "have not" state, local, county, and tribal governments with both human and technical capacity. NSGIC (National States Geographic Information Council) *representing all 53 states and territories is the recognized coordinating organization with the USGS and has become an effective body for raising congressional awareness about the importance of geographic data. Congress now provides a line item to maintain geographic data as part of the budget for the Department of Homeland Security. At the same time, a national fee for wireless communication has funded a high-resolution DEM (Digital Elevation Model) and the federal government is looking for new places to invest these funds. Federal agencies have minimal involvement in primary data capture. They serve as the quality assurance/quality control organizations.*

The private sector has developed a robust market for location-based services and is providing a wide range of spatial data services to local governments, therefore companies are amenable to more liberal licensing agreements. Local governments recognize that the benefits of freely distributing their data outweigh the minimal revenue they earn. County planning offices generate substantial business opportunities due to frequent updating requirements. Graduation from high school requires attainment of geographic literacy standards. Several universities now offer courses in the NSDI. The libraries are infused with new funding for large-format paper-map reproduction, digital and hardcopy archiving, and organization of metadata. Researchers have provided the disabled with embedded assistive technology from NSDI to support personal navigation. Finally, to prepare for the 2010 census, the Bureau of the Census simply waits until January 2010 and downloads the current version of the enhanced National Atlas

Although perhaps optimistic (e.g., completed Framework layers, smooth workflow processes, availability of financial resources) the scenario articulates some of the changes that will be necessary if *The National Map* is to move from a concept to a reality.

CHANGING ROLES, FAMILIAR ISSUES

The activities of the USGS's National Mapping program have changed significantly over the last 30 years, from a role of being a producer and maintainer of paper map products (principally the 1:24,000 seven-and-a-half-minute series) to that of developing digital core geographic databases through collaborative programs. Demand for federal digital map products has soared (Lemen, 1999). At the same time, the agency has suffered significant real budget reductions (e.g., USGS, 2002a), making it difficult for them to play the new role as catalyst for multipartner program collaboration. The cuts also caused the agency to scale back the updating of the 1:24,000-scale map series. Increasingly the USGS has out-sourced such map tasks as orthophotography acquisition and maintenance.

The committee reviewed previous evaluations of federal mapping programs to build out the NSDI (see Appendix C). These programs have often been led by USGS. The committee was struck by the fact that in spite of over 20 years of largely convergent statements of clearly identified needs and strategies, a "national map" as envisioned in the preceding scenario has not been forthcoming. Numerous partially successful spatial data initiatives have come about that have in various degrees completed some of the parts of what should be considered a national map. This has happened largely in spite of, not because of, the various attempts at coordination at the federal level.

Fortunately the Office of Management and Budget Circular No. A-16 is unambiguous about the tasks federal agencies should perform (see Box 2.1) and the key role the FGDC should play in coordinating federal mapping. Given the critical earlier role of U.S. federal data in the spatial data community and in establishing world leadership in this nascent industry, the committee believes this federal role is also one of creating and maintaining a map of the nation for the public domain. The committee recognizes that the intermediate-scale (1:250,000 to 1:24,000), nationally consistent paper maps of the past are a poor future model. The opportunity to use existing local agency data creatively to provide higher resolutions, more detailed map scale, and timely map coverage should not be missed.

BOX 2.1
USGS Roles Described in the Office of Management and Budget Circular No. A-16, Revised August 19, 2002

Since 1953 with revisions in 1967, 1990, and 2002, Circular No. A-16 has provided direction for federal agencies that produce, maintain, or use spatial data either directly or indirectly in the fulfillment of their mission. The circular establishes a coordinated approach to development of an electronic National Spatial Data Infrastructure and establishes the Federal Geographic Data Committee (FGDC) (see Box 1.3). Certain federal agencies have lead responsibilities for coordinating the national coverage and stewardship of specific spatial data themes. The roles of lead federal agencies include facilitating the development and implementation of needed FGDC standards and a plan for nationwide population of each data theme.

The data themes led by the USGS are (in alphabetical order) as follows.

BOX 2.1 Continued

Biological Resources. This dataset covers data pertaining to or descriptive of (nonhuman) biological resources and their distributions and habitats, including data at the suborganismal (genetics, physiology, anatomy), organismal (subspecies, species, systematics), and ecological (populations, communities, ecosystems, biomes) levels.

Digital Orthoimagery. This dataset contains georeferenced images of the Earth's surface, collected by a sensor in which image object displacement has been removed for sensor distortions and orientation, and terrain relief.

Earth Cover. This theme uses a hierarchical classification system based on observable form and structure, as opposed to function or use. The theme differs from the vegetation and wetlands themes, which provide additional detail.

Elevation (Terrestrial). These data are georeferenced digital representations of terrestrial surfaces, natural or engineered, which describe vertical position above or below a datum surface.

Geographic Names. This dataset contains data or information on geographic place names deemed official for federal use by the U.S. Board on Geographic Names.

Geologic. The geologic spatial data theme pertains to all geologic mapping information and related geoscience spatial data (geophysical, geochemical, geochronologic, paleontologic) that can contribute to the National Geologic Map database.

Hydrography. This data theme comprises such surface water features as lakes, ponds, streams and rivers, canals, oceans, and coastlines.

Watershed Boundaries (co-leaders: Department of the Interior, USGS; and U.S. Department of Agriculture, Natural Resources and Conservation Service). This data theme encodes hydrologic, watershed boundaries into topographically defined sets of drainage areas.

SOURCE: OMB (2002).

BENEFITS AND CHALLENGES OF SHARING DATA

One goal set forth by the USGS (USGS, 2001) is to seek the most reliable, accurate, and timely data and use it to aggregate information up to a consistent national coverage. Prior national mapping efforts have created essentially stand-alone maps at specific scales, such as 1:100,000 and 1:24,000, whose update and production are independent operations. Although this was necessarily true in a paper map era, far more is possible in the digital era. Best available data are often held by the private sector and are proprietary, or at the local, municipal, or county government level and are made available at significant cost or with restrictions on use. The problem becomes one of how these data can be "contributed" to a broader database, such as *The National Map*. The greatest challenges will be coordination among hundreds of participants and developing incentives for state and local governments to share and standardize their data or metadata (e.g., see comments of Scott Cameron, Appendix D). The greatest benefits will be an enrichment of the entire national coverage.

To achieve data sharing through the Mapping Partnership program (USGS, 2002b) the USGS has in the past used cooperative partnerships in the forms of conventional, innovative, and framework partnerships, and Cooperative Research and Development Agreements (CRADAs). Partnerships have obvious benefits. For example, given a dual need for map revision and digital orthophotographic map production at the federal level, and the desire for a high-precision spatial database for facilities planning, local taxation, or even individual projects at the local level, the committee believes that the best way to conduct this mapping is by joint funding and shared commitments to acquisition support, map data processing, and data production. The committee believes that local strengths are in site-specific knowledge, transactional updates, programmatic requirements, and in utility; federal strengths are in coordination, support, development and application of standards, and data archiving and distribution. When the public domain is supported, local, state, and federal needs are met simultaneously and a powerful new model for public data sharing presents itself. Locally acquired data, including updates, could find their way into federally administered mapping programs and coarser-scale applications that suit national needs. In such an organizational circumstance all of the parties involved reap benefits.

BLANKETS AND QUILTS

The committee's discussions covered two approaches to national mapping. On the one hand there is a need for a national map that is complete, consistent, at a common scale, and that can be updated selectively. A metaphor for this approach is a blanket, which covers a user uniformly at a single weight but is available universally at minimal cost. Earlier approaches that resulted in "blankets" are the 1:24,000 and 1:100,000 topographic map coverages, the Bureau of the Census's TIGER files, and the Digital Orthophoto Quadrangle (DOQ) map at 1:12,000 equivalent scale and 1-meter spatial resolution.[1]

An alternative approach follows the patchwork quilt metaphor. Here, individual squares are created by many processes (weaving, knitting, crochet), to different specifications, and for varied purposes, resulting in different materials and colors or fabric. In the metaphor the different weaves correspond to different map scales. Put another way, the different patches correspond to different levels of data resolution.

For individual pieces of the quilt to fit together there must be agreement on what size square to use, and on how the square patches will be assembled. For the creators willing to accept an overall set of design specifications, the benefit is in getting use of the quilt and in seeing their work in context. To further stretch the metaphor, the maintenance of each square could be the responsibility of its creator, thus creating a self-maintaining quilt.

The USGS vision for *The National Map* contains blanket and quilt components. The blanket component is the proposed continuation of the 1:24,000 series maps (USGS, 2001, p. 15), the proposed seamless national coverage with Landsat imagery (USGS, 2001, p. 10), and the proposed incorporation of the National Atlas into *The National Map* (USGS, 2001, p. 17). The quilt component is evident in the proposed high-resolution patches of data where local or state partners link their higher-resolution data (USGS, 2001, p. 2).

The importance of a quilt-first or bottom-up component to *The National Map* cannot be overstated. The blanket approach to national mapping has in the past been exclusively a top-down operation. As the USGS has built state and local cooperative arrangements as parts of cost sharing and cooperative research agreements, the blanket-only model has become increasingly strained. The most effective arrangements for federal collaboration are those in which both parties have a mutual interest in advancing the goals of accurate and timely mapping. The National Digital Orthophoto program (NDOP) and TIGER modernization programs are notable examples (see Boxes 2.2 and 2.3). Other examples

---

[1]Recognize that these blankets did not form overnight. In essence they become blankets when the last quadrangles are completed. What differentiates these from the patchwork quilt model is that a uniform scale was used from the outset. The blanket and quilt metaphor cannot be pushed indefinitely: Even a blanket with a uniform scale, by the nature of its construction, will have variable currency and accuracy. Nevertheless, differences in scale, accuracy, and currency can be supported at the distributed (quilt) level, whereas the blanket plays the role of being consistent, seamless, and uniform.

of partnerships that serve mutual interests are the Natural Resources Conservation Service's (NRCS) National Watershed Boundary dataset project, the U.S. Department of Agriculture's National GIS (Geographic Information System) Implementation plan, Bureau of Land Management's Land Use plans, and the National Biological Information Infrastructure (NBII).[2] Additionally, lessons can be learned from the bottom-up organIzational component of *The National Map* pilot projects[3] (see Table 2.1).

Central to the success of the patchwork quilt model is that detailed mapping by a local agency or company can likely have value and quality added by being a part of *The National Map*. Federal mapping efforts also benefit from the continued reporting of modifications, updates, and new information that come from the local level. Models are available that favor exchanges based on incentives other than money, for example, exchanges based on the concept of equivalent value. Since updates are essentially the result of local processes, local reporting of updates–a sort of parallel processing for the nation–represents a way that a national map can be up-to-date, or at least sufficiently so to meet national needs, such as homeland security and natural disaster response. Acceptance of data into *The National Map* itself could have value. Similar to a "National Historical Landmark" designation, contributors would have their data certified as being from a "*National Map* partner." Such designation could have implications for future data sharing, future cost sharing, and ongoing relationships associated with NSDI activities, and is more comprehensive and realistic than simply complying with fixed National Map Accuracy Standards. For example, a *National Map* partner county could be formally assured by the USGS that their census block and tract delineations conformed to locally maintained street and curbline map databases. When data are returned to participating cities and counties with increased value, the effect would also benefit the surrounding regions served by the city or county. When two adjacent counties conduct their mapping, neither has a mandate to ensure that a mapped road exiting one county arrives in the

---

[2]The NBII provides a concrete example of how USGS manages partnerships with multiple entities. The NBII is a collaborative program to increase access to data and information about the nation's biological resources. It links databases, information products, and analytical tools maintained by NBII partners and other contributors in government agencies, academic institutions, nongovernment organizations, and private industry (USGS, 2002c).

[3]There are numerous technical and institutional issues that need to be addressed to implement *The National Map*. Nine pilot projects are underway to identify some of the key issues that may arise, to develop business plans, and to develop technical solutions.

correct place in the next county. If the USGS took responsibility for this kind of integration, both counties would benefit by participating in the project because each would get back a map showing their roads in the same position, thus connecting with each other. Ultimately, *The National Map* must be easy to understand, easy to participate in, and have obvious benefits for all stakeholders (e.g., see comments of Dennis Goreham, Appendix D).

BOX 2.2
The National Digital Orthophoto Program

Digital orthophotos are computer-compatible representations of aerial photographs with displacements and distortions caused by terrain relief, atmospheric conditions, and camera systems removed. The initial proposal for a National Digital Orthophoto program (NDOP) was made by the NRCS (formerly Soil Conservation Service) in May 1990. Primary support for the program came from the NRCS, the Agricultural Stabilization and Conservation Service (ASCS), and the USGS. The NDOP was funded cooperatively to provide national coverage in five years, with a 10-year average update cycle. Production for the estimated 220,000 digital orthophoto quadrangles (DOQs) in the conterminous United States used the capability of the USGS and commercial contractors, the majority of work being done by the private sector.

The federal land and resource management agencies are the principal users of DOQs, but many state and local governments and utilities with extensive geographic information system (GIS) operations are at the leading edge of innovation in digital orthophoto applications. Applications include land and timber management, routing and habitat analysis, environmental impact assessment, evacuation planning, flood analysis, soil erosion assessment, facility management, and groundwater and watershed analysis.

All DOQs are referenced to the North American Datum of 1983 (NAD 83). USGS DOQs must meet National Map Accuracy Standards for their respective scale map quadrangles. The specifications for the digital orthophotos are designed to accommodate DOQs from various producers. Published by the USGS as the Standard for Digital Orthophoto Quadrangles, they have been endorsed by the FGDC. Participation in the NDOP can occur through data-share, work-share, and joint-funding agreements. The results must meet the FGDC DOQ standards.

The USGS Innovative Partnership program was developed to acquire digital spatial data that could be made available in the public domain from nonfederal sources. In October 1993 the list of Innovative

Partnership data types was expanded to include DOQs. The program initially aimed at complete coverage in five years but was too ambitious given the federal budget situation. However, the rapid growth of state and local government interest and participation helped the program to near completion of first-time coverage of the nation in 2002.

The NDOP partnerships have proven essential among all levels of the spatial data community, and serve as a good model for the nation in the development of other digital base map data.

BOX 2.3
TIGER as a Role Model

The entire nation was mapped to the block level for the first time for the 1990 decennial census. Although mapping by the Bureau of the Census was automated long before that, it was the TIGER line files of the 1990 census and the ability to disseminate these on CD-ROM that enabled this array of information to become widely available to the public for the first time. Now the bureau has signed a memorandum of understanding with the USGS to become a partner on street centerline data in *The National Map* (Robert Marx, Bureau of the Census, personal communication, 2002).

Over the past two decades the Bureau of the Census has established a set of procedures and institutional arrangements for the creation, maintenance, and use of its TIGER line files. The representation of street centerlines and the associated address ranges are the key to a successful decennial census. The original version of the TIGER line files evolved during the 1980s through a cooperative agreement with the USGS. The addresses are compiled in the Master Address File (MAF), which is continually updated and becomes the Decennial Master Address File for the decennial census. The MAF is maintained collaboratively by the U.S. Postal Service, local governments, and the private sector. The Postal Service provides the bureau with its Delivery Sequence Files, which are tested for geocoding against the TIGER files to determine whether each address can be assigned to a unique census block. Through its 15 regional offices the bureau conducts a canvass of every block. Concurrently the MAF can be shared with local government units through the Local Update of Census Address (LUCA). The LUCA program provides the local government with an opportunity to identify addresses that should be added or deleted. These suggestions are field verified by the bureau. The bureau also has an arbitration process in place that includes independent review for any appeals that a local government may have to the

BOX 2.3 Continued

MAF. As census day approaches the New Construction operation enables local government entities to identify new housing units. The private sector also plays an important role in the TIGER maintenance program. These arrangements vary from agreements with Geographic Data Technology to provide streets, street names, and address ranges to TIGER, with First Data Solutions to provide residential address lists, and with the Environmental Systems Research Institute (ESRI) to develop user-friendly software to manipulate features on maps. The creation of the American Community Survey (ACS) will provide an opportunity for continuous update of the MAF throughout the decade. The ACS program aims to provide "accurate up-to-date profiles of America's communities every year" (Bureau of the Census, 2002a). When the ACS becomes operational in 2003 it will rely on a "self-enumeration through mail-out/mail-back methodology." This method has procedures for telephone and field verification of the addresses of surveys not returned and will enable the bureau to update, maintain, and verify its address files on a continuing basis.

The Federal Depository Library program (FDLP) of the U.S. Government Printing Office made the TIGER files and the census summary tape files for the 1990 census available to more than 1,300 libraries. Libraries are depositories for census information from the first census in 1790 forward; the electronic distribution of these data opened up new opportunities for libraries to provide access for analysis and interpretation over any desired geographic area.

TIGER files and advances in technology have increased the speed with which GIS has become a multidisciplinary tool for interpreting and analyzing data of many types. By putting these data in the FDLP the Bureau of the Census demonstrated that public domain data can provide opportunities to the commercial sector for product development. At the same time the citizenry, including researchers, can benefit from government information in the public domain. A case in point is the participation of many of the FDLP libraries in the 1992 GIS Literacy project designed by the Association of Research Libraries working with ESRI. Workshops were conducted around the country for librarians to learn ArcView and identify software needs in the library community. Not only did hundreds of librarians become GIS literate but the commercial sector also learned how it could develop value-added products using the TIGER files and census data. The knowledge of GIS and its potential spread from libraries to K-12 programs in the nation's schools.

SOURCE: Bureau of the Census (2002a,b,c)

The patchwork quilt approach for *The National Map* could meet the ambitious update goals set by the USGS. This type of national map would need to be flexible, be controlled autonomously rather than centrally, contain multiple spatial resolutions, be dividable spatially in almost any way (e.g., by census tracts, counties, congressional districts, watersheds, health districts), and would require a degree of public-private cooperation that is unprecedented (e.g., see comments of Hugh Bender, Appendix D). This patchwork quilt approach is bottom-up. If *The National Map* were built on this approach, the role of the USGS would be to act as an initiator of new working collaboratives, as a nurturer of collaboratives, as a coordinator, and as the guardian of consistency, standards, quality, and reliability. This role involves policing and facilitating content derived from the best available data contributed from local sources. It is separate from the provision of access tools enabling a user to view and utilize data. Many agencies, public, private and academic, will seek to provide portals to such a blanket and quilt system. Some systems are already based on this model, such as ESRI's Geography Network and Geographic Data Technology's Community Update.

TABLE 2.1 *National Map* Pilot Projects

Project/Location	Focus Areas
Delaware	▪ Current, seamless base geographic datasets for entire state; ▪ Partnerships with multiple agencies in Delaware; ▪ Website addressing data integration issues: <http://www.datamil.udel.edu/nationalmappilot>.
Lake Tahoe, California-Nevada Area	▪ Integration of federal and local data; ▪ Use of volunteer groups for updating datasets; ▪ Collaboration with California on statewide vision for Framework data development and distribution.
Mecklenburg County, North Carolina	▪ Demonstration of current capabilities to provide access to, combine, view, and download a wide range of data themes from distributed data holdings at the national, state, and county levels; ▪ Addressing such homeland security issues as continual service through mirror and backup data sites, and reprojecting data "on the fly."

TABLE 2.1 Continued

Project/Location	Focus Areas
Missouri	▪ Multihazards risk assessment, mitigation, and emergency planning in southeastern Missouri; ▪ Incorporation of best available data by working with local agencies; ▪ Integration of geologic data with *The National Map*.
Pennsylvania	▪ Orthoimagery with frequent updates at county level to demonstrate data currency, archival, and maintenance functions of *The National Map;* ▪ Hydrography data coordinated with local sources; ▪ Land-cover data coordinated with state and local agencies.
Texas	▪ Expansion of USGS relationship with StratMap (Texas Natural Resources Information System database) as primary data source to *The National Map*; ▪ Prototype map product and digital output mechanisms; ▪ Refining local data linkage and integration techniques; ▪ Website: <http://www.tnris.state.tx.us/digitalquad/>.
Utah	▪ Testing USGS capabilities to support critical transportation data for emergency response; ▪ Urban and rural testbeds; ▪ Integration of local transportation data.
Washington-Idaho	▪ Creation of new partnerships in rural areas lacking solid geospatial information; ▪ Multijurisdictional, interstate project.
U.S. Landsat	▪ Providing access to Landsat satellite imagery as reference layer for *The National Map*; ▪ Dynamic update cycle with archival and records retention; ▪ Website: <http://gisdata.usgs.net/website/landsat>.

SOURCE: USGS (2002d).

A combination approach of a blanket and quilt (see Figure 2.1) would ensure that a broad functional database was available nationally (the blanket), with

- bottom-up feature-level[4] updates;
- enforced consistency;
- a single scale (of, say, 1:12,000 or 1:24,000);
- consistent generalization to alternative spatial reference frames (e.g., such spatial units or geographic extent as congressional or police districts); and
- access over the Internet, with instantaneous availability for any citizen, agency, or community group.

At the same time, the patchwork quilt would contain public domain data and pointers to locally held proprietary data where such restrictions are necessary. Proprietary data could take the forms of (1) public domain metadata that includes contact and purchase or access information; (2) watermarked or encrypted data sets accessible for browsing but for download only by use of a controlled-access key; (3) degraded thumbnail images for browsing and limited use; or (4) fully public domain data. Any entity could contribute a data theme, but only those data used for updates in the blanket and that pass the acceptance tests of the USGS would receive *The National Map* status or endorsement.

Although this vision is somewhat consistent with that outlined by the USGS vision document (USGS, 2001), the facilitation and systems integration role for the USGS is a radical departure from prior activity. The success of integration efforts and later data update arrangements will depend on the ability to leverage local government, private, academic, and nonprofit support as a new kind of collaborative in a new kind of federal role. USGS is suited to take on this initiative because it has the experience, because it has a mandate in OMB Circular No. A-16, and because it already has part of the infrastructure (field offices and a spatially distributed system of centers) that is needed for success. The USGS support can come in several forms: as a broker of funding, a provider of technical expertise and advice, a provider of technical geospatial resources, and a coordinator of development efforts at all levels.

---

[4]Feature-level means at the hierarchical step corresponding to a single mapped object, such as a street segment, pond outline, or geodetic benchmark.

[5]Feature-level means at the hierarchical step corresponding to a single mapped object, such as a street segment, pond outline, or geodetic benchmark.

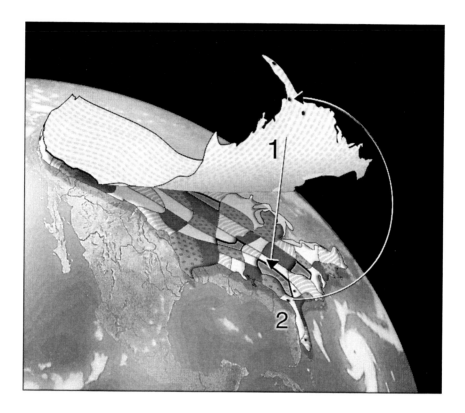

FIGURE 2.1 Illustration of the blanket and quilt metaphor and how the
two components complement each other. The blanket covers holes in the
quilt (1), and the quilt provides data and updates to the blanket (2).
SOURCE: Susanna Baumgart, University of California, Santa Barbara.

SUMMARY

A national map as conceptualized by the USGS is justified and
timely. Authority for such an enterprise is already delegated, yet the
undertaking will require a cultural shift in the way that the USGS
approaches its mapping. The future of mapping in the United States
depends on forging partnerships among all levels of government and the
private sector.

3

Components of a National Map

INTRODUCTION

This chapter outlines how the USGS *National Map* concept might gain wider understanding, acceptance, and participation by building on an existing successful program, the National Atlas. The chapter then reviews the proposed data layers of *The National Map* in the context of other federal programs, and ends by examining the proposed data layer characteristics.

A CONTEXT FOR *THE NATIONAL MAP*

Continuing the blanket and quilt metaphor introduced in Chapter 2, the committee sees a national map with three interdependent, hierarchical components. A blanket map coverage for the nation is the top tier, a patchwork quilt is at the next tier, and any data that are not formally part of the other two levels occupy the third tier. The first two tiers form the core of the NSDI (see Figure 3.1).

The defining property of the blanket tier is a continuous spatial coverage at standard map scales and a common set of features and themes focused on the Framework data layers identified by the FGDC. We term this the enhanced National Atlas, borrowing the name from and expanding upon an existing program (see Box 3.1) to distinguish it from

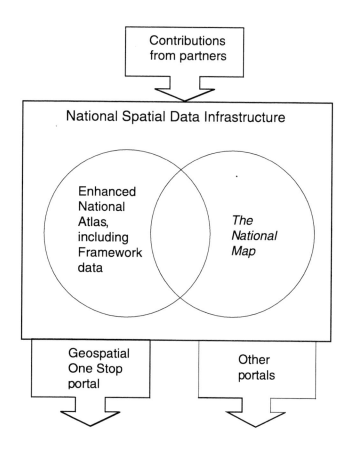

FIGURE 3.1 Relationship of the National Spatial Data Infrastructure (NSDI), the enhanced National Atlas, *The National Map*, and other components. The NSDI provides the umbrella of data standards, prototcols, partnership arrangements, interagency cooperation, and data. Within this lie the enhanced National Atlas and *The National Map.* Their overlap represents the processing (generalization) of the large-scale information into the enhanced National Atlas and the filling of "holes" in the patchwork quilt *National Map* with atlas data when no local data exist.

BOX 3.1
Existing National Atlas of the United States

The USGS has been publishing National Atlas products since 1970 (see <http://nationalatlas.gov>). Intended for use at national or large regional scales, most of the map layers in the National Atlas are compiled at a scale of 1:2 million and cover the full geographic extent of the United States (although some cover the conterminous United States only). Many federal agencies supply data to the National Atlas, and linkages are provided to agency sources where their information is portrayed.

The National Atlas includes the following data services:

Make Maps, Online Interactive Maps. Available since 1998.

Multimedia Maps. Multimedia technologies are used to supplement and enhance cartographic information products; to tell narrative stories; to deliver interactive maps; to develop maps that show change over time; and to facilitate the visual presentation and understanding of complex geographic phenomena and relationships.

Map Services. Over 500 integrated map layers are accessible as two types of Web mapping services. First, Open GIS Consortium-compliant Web mapping services are offered so that developers can embed the National Atlas in their applications, and second, the National Atlas publishes map services through ESRI's (Environmental Systems Research Institute) Geography Network.

Printed Maps. High-quality paper maps depicting a variety of national conditions.

Printable Maps. Page-size (8.5 by 11 inch) maps designed for printing and reproduction.

Spatial Databases. A variety of information and statistics with a spatial component from federal partners.

Metadata. Metadata files include information describing the contents of each map layer, such as how it was made, its lineage, and its quality.

Scientific and Expository Articles. Cartographic products are supplemented with narratives.

In addition to the National Atlas site several other federal agencies offer similar data or even the same data with different data delivery engines and services. For example demographic information can be viewed, queried, and used to build custom maps on the U.S. Bureau of the Census website and the Bureau of Land Management's Geocommunicator offers a perspective on federal ownership.

the patchwork quilt of *The National Map*.[1] The primary differences between the current National Atlas and the enhanced version are scale and theme. The themes would include NSDI Framework data: cadastral, digital orthoimagery, elevation, geodetic control, hydrography, political and administrative boundaries, and transportation. To these would be added geographic names and land use. The equivalent map scales would be those already in use for the National Mapping program, with 1:24,000 (1:12,000 for orthoimagery) being the finest.

Changing the terminology introduced by the USGS is desirable because without a distinction between the two different types of information it will be difficult to sell the *National Map* concept at the level of the local data contributor. With an enhanced National Atlas there is more task separation: The atlas provides a comprehensive, up-to-date, and uniform map of the nation. By dividing off the quilt from the blanket, at least conceptually, clearer federal responsibilities are evident, and *The National Map* can become a more participatory and collaborative enterprise. In practice the atlas and map would be intricately linked. In the absence of local participation the spatial content of *The National Map* would be that of the National Atlas. With participation the enhanced National Atlas would contain the latest updates at specific scales from *The National Map*.

An Enhanced National Atlas

An enhanced National Atlas could constitute a nationally complete dataset including the Framework data layers as specified for the NSDI, at a series of such scales as 1:24,000 (or perhaps 1:12,000), 1:100,000, 1:1 million, and smaller. The initial contents of the enhanced atlas would be those outlined by the USGS (i.e., geographic names, topography, orthophotography, land cover, hydrography, transportation, structures, and boundaries). To fulfill the needs of the federal mapping community (e.g., BLM, Census, EPA, NASA, NIMA, NOAA, USDA, USGS, and others), it seems of high value and low cost to add coastline data from NOAA, detailed data on the Public Lands Survey System (PLSS), and cadastral data on public lands, geodetic control, and surface geology. Although collectively this is beyond the scope of the USGS topographic mapping responsibility, such a federal collaborative would have many benefits, and continues the spirit of the 1884 National Academy of Sciences report (NAS, 1884) by reducing duplication and increasing collaboration. The

---

[1]The USGS vision document (USGS, 2001) indicates that the National Atlas is a component of *The National Map* but does not expand on their relationship.

data would be updated with each new change through collaborations between USGS and its partners and forwarded to a state or regional integration office transaction by transaction.

The USGS would be responsible for metadata provision and support, data archiving, and data integration in their role as stewards of the enhanced National Atlas. The first pass could consist of larger-scale Framework data at the 1:12,000 or 1:24,000 scale such as geodetic control, digital orthophotos, PLSS data, and public ownership boundaries. Other data types, such as hydrography and transportation, may not be completed for several years since they require significant integration with other data types at the 1:12,000 or 1:24,000 scale.

The Committee's View of *The National Map*

The committee sees *The National Map* as a suite of data at multiple scales, with variable spatial extent, contributed by *National Map* partners. Themes would be as in the enhanced National Atlas, but local interpretation and adaptation to scales and conditions would be permitted. Some entries in *The National Map* would be only metadata, others browse-only thumbnails; and most would be public and private data voluntarily placed in the public domain. The data in *The National Map* would be accessible through Geospatial One-Stop and other portals, many at federal, state, and local agencies and businesses. The USGS would evaluate and integrate (e.g., reclassify, merge, and mosaic) the data to be consistent with adjacent datasets, and serve as the guarantor of completeness, accuracy and consistency.[2] This could be done by a "seal of approval" technique that would go beyond arbitrary paper map standards such as the National Map Accuracy Standard, and instead be modeled on the truth-in-labeling approach used by the Spatial Data Transfer Standard. The basis of certification and metadata would be the individual feature, not the tiling or coverage system. Accepted and certified data and updates would be passed into the enhanced National Atlas and back to the data custodians at the local level. Any patches not filled by local data would be filled by smaller-scale data from the atlas until new local data are collected.

The enhanced National Atlas and *The National Map* would be core datasets in the NSDI. Data that did not meet *National Map* standards would reside in the broader NSDI, as they do today. These data could be

---

[2]The USGS already has experience of the guarantor role through the National Digital Orthophoto Program (see Box 2-2), however the challenge of fulfilling this role for many data themes will be great.

accessed through a portal of portals, passing the user between distributed sources as necessary. Existing data portals such as Destination: Earth, Gateway to Earth, GeoExplorer, Geography Network, Geospatial One-Stop, and GISDATA Web Mapping Portal could be seen as entry points into the U.S. component of the Global Spatial Data Infrastructure.

PAPER MAPS

The continued production of USGS paper map quadrangles using standard lithographic presses places demands upon the USGS that are inconsistent with a new role as facilitator and custodian of the proposed *National Map* and atlas. Consequently, new distribution mechanisms are needed. Because of the different user groups, the distribution mechanisms for *The National Map* and enhanced National Atlas would not completely overlap. *The National Map* would be distributed on the Internet. The content of the enhanced National Atlas could continue to serve the needs of paper-map users through Internet-linked print-on-demand approaches, yet add new and flexible options.

The USGS goal of liberating the printed map from its spatial reference frame is probably the most appealing to users of hardcopy maps, and there will undoubtedly be a continued demand for paper maps (e.g., see comments of John Voycik, Appendix D, and NRCan, 2002). The flexibility of map printing at arbitrary scale, extent, and theme, and at the user's convenience is also attractive and could increase the use and demand for map products. Users should be able to print from the Internet, but should also have access to high-quality hardcopy. Barring any restrictions on the file size of downloading capability, FDLP libraries can provide full-size 1:24,000 (24 x 30 inch) color prints. Map machines at USGS Earth Science Information Centers and at the National Geographic Society can also provide an entire topographic map on a 13 x 18-inch sheet at scale of about 1:32,000 (through a collaborative project between USGS and the National Geographic Society). These are good examples of what is possible.

The continued production of hardcopy map products is probably a commercial opportunity best exploited by creating USGS partnerships based on the content of the enhanced National Atlas. Commercial map producers have already taken advantage of the accessibility of public domain USGS and other digital map data, such as TIGER files. A healthy map production sector is in the public interest, and also could meet federal needs.

PROPOSED DATA CONTENT

The USGS proposes that its *National Map* will include the following NSDI Framework data themes (USGS, 2001):

- Orthoimagery (high resolution) (USGS lead designated in OMB Circular No. A-16, and an NSDI Framework layer);
- Elevation and bathymetry (high resolution) (A-16 lead on elevation);
- Hydrography (A-16 lead);
- Transportation (roads, railroads,waterways);
- Administrative boundaries; and
- Publicly owned lands.

Three non-Framework layers also are included:

1. Geographic names (e.g., physical and cultural features) (A-16 lead);
2. Land cover; and
3. Structures.

The USGS vision document also lists potential partnerships to link with such data as geodetic control and bathymetry (both are NOAA responsibilities).

The USGS themes are now examined in the context of two other federally organized spatial data activities with which the USGS project will need to dovetail (not forgetting of course that coordination with state and local activities provides additional integration challenges for the USGS as it pursues its mapping vision).

The Relationship of the *National Map* Concept to Activities of Subcommittees of the Federal Geographic Data Committee

The Federal Geographic Data Committee has established several thematic subcommittees (see Figure 3.2). These subcommittees are charged with coordination of federal data efforts and developing standards for each theme. The focus of each of these subcommittees is compared to the proposed data content of the USGS's *National Map* in Table 3.1. A number of the subcommittees are now discussed in more detail to illustrate the overlap with *National Map* themes.

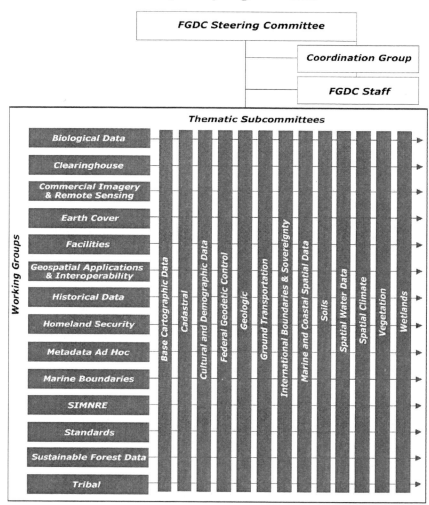

FIGURE 3.2 Federal Geographic Data Committee organization. SIMNRE stands for Sample Inventory and Monitoring of Natural Resources and the Environment. SOURCE: FGDC (2002d).

TABLE 3.1 Comparison of Focus of FGDC Subcommittees and
Proposed *National Map* Content

FGDC Subcommittees and Lead Department	*National Map* Content
Base Cartographic, Department of the Interior	Orthorectified imagery (high resolution) Elevation (high resolution) Administrative boundaries
Cadastral, Department of the Interior	Publicly owned lands
Cultural and Demographic, Department of Commerce	Administrative boundaries Cultural features (geographic names)[a]
Geodetic, Department of Commerce	– [b]
Geologic, Department of the Interior	–
Ground Transportation, Department of Transportation	Transportation (roads, railways, waterways)
International Boundaries and Sovereignty, Department of State	–
Marine and Coastal Spatial Data, Department of Commerce	Bathymetry[b]
Soils, Department of Agriculture	–
Spatial Water Data, Department of the Interior	Hydrography
Spatial Climate, Department of Agriculture	–
Vegetation, Department of Agriculture	Land cover[c]
Wetlands, Department of the Interior	Land cover

[a]The U.S. Board on Geographic Names is also a source for geographic names.
[b]A partnership with NOAA is proposed in the USGS vision document to provide a link with this data content.
[c]FGDC also has an Earth Cover Working Group that deals with land cover.

Subcommittee on Base Cartographic Data

The orthorectified imagery, elevation and bathymetry, and administrative boundaries are components of the base cartographic subcommittee
chaired by the Department of the Interior. According to the charter of
the Subcommittee on Base Cartographic Data (FGDC, 2002e) it has
"responsibility to coordinate base cartographic data-related activities as
assigned to the Department of the Interior." The activities of the
Subcommittee on Base Cartographic Data include standards development
in orthoimagery, elevation, positioning, and accuracy. In the context of
the USGS's proposed *National Map* three provisions of the charter are
noteworthy.

1. Determine which categories of base cartographic data are to be
included in the National Digital Cartographic Data Base[3] and recommend
the addition of other categories of base cartographic data not currently
being collected.

2. Assist the USGS in establishing and publishing standards and
specifications for the data (for example, incorporating conversion to
metric units), and assist in establishing priorities for base cartographic
data production.

3. Establish and maintain mechanisms for interface among databases
of agencies participating in the National Geographic Data System.

Arguably, the USGS *National Map* vision includes data types and functions (such as standards development) that are the responsibilities of this
subcommittee. The data content that falls into the domain of the Subcommittee on Base Cartographic Data is as follows:

> Base cartographic data is defined as the fundamental data set of geographic
> data that is normally produced in the preparation of national series general
> purpose graphic and digital cartographic products. These data represent the
> physical and cultural features (natural, artificial, or both) of part or the
> whole of the Earth's surface. Used individually or collectively the data can
> provide the framework upon which other themes of geographic data can

---

[3]The National Digital Cartographic Data Base is a term often applied to the
entire public domain digital map holdings at the USGS EROS (Earth Resources
Observation Systems) data center. It includes most of the data that would
become the enhanced National Atlas.

be referenced. Examples of these cartographic data include roads, streams, contours, and geographic and other coordinate reference systems.

Base cartographic features are essentially the standard layers of the USGS seven-and-a-half minute topographic series, which consists of eight overlays.

1. Boundaries;
2. Hypsography (contours);
3. Hydrography;
4. Miscellaneous Culture;
5. Nonvegetation;
6. Vegetation;
7. Geodetic Control; and
8. Transportation.

The *National Map* content will therefore be familiar to paper and standard USGS digital map product users, promoting continuity. These layers also are multiuse and fundamental to the location of many other objects in different layers. Current USGS map databases that include these themes are available through USGS's U.S. GeoData program: digital elevation models, digital orthophoto quads, digital line graphs, and digital raster graphics.

Subcommittee on Cadastral Data

The only mention of cadastral data in the *National Map* vision is with respect to publicly owned lands (USGS, 2001). It is logical for the Bureau of Land Management, which directs the FGDC Subcommittee on Cadastral Data, to have *The National Map* include its property records in the public domain states. For example, inclusion of the Public Land Survey corners would be an important step in this process. Although it would be beyond the scope of a federal program to include the parcel boundaries of a local cadastre, the value of *The National Map* would be enhanced by having street centerlines with sufficient positional accuracy to interoperate cleanly with local parcel boundary and centroid datasets. If local government cannot insert its high-resolution parcel-level data, there will be at least two competing versions of base maps for all urban areas of the United States. That would be a waste of time and effort. One driver behind improved cadastral data is from the planning community, which has a growing need for consistent digital cadastral data across political jurisdictions. Another is local taxation and assessment of real property. At the state level, public lands and holdings are important. In general it is in the national interest that cadastral boundaries of state, local, and federal lands and facilities be available. The USGS has not clearly specified the scope of features that partners and USGS would like included in *The National Map*. It is important to specify such a scope while the project is still at the concept stage.

Subcommittee on Cultural and Demographic Data

The focus of the FGDC Subcommittee on Cultural and Demographic Data, chaired by the Department of Commerce, translates most directly to the *National Map* categories of government unit boundaries and cultural features. This subcommittee focuses on data that include the character-istics of people, the nature of dwellings in which they live, the economic activities they pursue, the facilities they use to support their recreational and health needs, the environmental consequences of their presence, and the boundaries, names, and numeric codes of geographic entities used to report the information collected (Bureau of the Census, 2003). These features are likely to be of major interest to *The National Map* users.

Subcommittee on Ground Transportation Data

The spatial representation of the center of streets is probably the most redundant data theme in the United States. Because of the large number of data collectors and stakeholders, a relationship between the FGDC Subcommittee on Ground Transportation Data and the USGS *National Map* project is critical and potentially complicated. The FGDC sub-committee is chaired by the Department of Transportation (DOT). This oversight is logical and provides a link to the DOT's state counterparts. However, the Bureau of the Census maintains the most complete and fully attributed nationwide digital representation of the road network. Since the Bureau of the Census is not a federal mapping agency its road network is really a work product of Census activities and has been designed neither for navigation nor optimized as a base map for large-scale local applications. Consequently, state DOTs and local highway agencies often develop their own versions of many road features for pavement management and facility inventory applications. At the same time, many local governments and private companies, aided by high-resolution imagery and Global Positioning System (GPS) technology, have developed files with a high degree of spatial accuracy that are used extensively for onboard vehicle navigation, in E911, and to support cadastral mapping. The potential for additional redundancy is increased further by the numerous proprietary and competing versions of street centerlines maintained by the private sector. The FGDC Subcommittee on Ground Transportation has developed a transportation feature standard that we hope will accommodate the needs of most users.

As part of TIGER modernization the Bureau of the Census needs to improve the positional accuracy of its street centerlines and is surveying every county to determine the best version of street centerlines. Ideally, high-quality county files will become part of TIGER and *The National Map*. A memorandum of understanding between Census and the USGS relating to sharing of street centerline data was signed on September 7, 2001 (Robert Marx, Bureau of the Census, personal communication, 2002).[4]

Marine and Coastal Spatial Data Subcommittee

The bathymetry theme in *The National Map* falls within the focus of the Marine and Coastal Spatial Data Subcommittee. Chaired by the Department of Commerce, this subcommittee promotes standards of accuracy and currentness in bathymetric and nautical charting data financed in whole or in part by federal funds. The subcommittee also exchanges information on technological improvements for collecting bathymetric and nautical charting data; encourages the federal and non-federal community to identify and adopt standards and specifications for bathymetric and nautical charting data; and collects and processes the requirements for federal and nonfederal organizations for bathymetric and nautical charting data (NOAA, 2002).

The Earth Cover Working Group

The USGS has a long history of conducting studies that deal with land cover. It is therefore appropriate that land cover be a part of their vision. Work on this component should be coordinated with the work of the FGDC's Earth Cover Working Group to establish consistent protocols

---

[4]In particular the Census will work with USGS "to develop processes to provide positionally accurate roads and updated boundaries of equal positional accuracy for use in *The National Map*. Other TIGER features–streams, lake/ river shorelines, railroads, pipelines, etc.–also will be provided to USGS in a manner to be jointly agreed upon as appropriate for both agencies."

and classification standards for land cover. The working group promotes an integrated, standardized, cost-efficient approach to identifying, classifying, and mapping features that cover the surface of the nation. The land cover component should also coordinate with the efforts of the FGDC Vegetation and Wetlands Subcommittees (FGDC, 2002a) to ensure consistency between the *National Map* land-cover theme and vegetation and wetland areas.

Framework Data

Over the past decade a consensus has evolved on how to build an integrated spatial database that will provide a consistent base for a wide range of applications. The construction process is analogous to constructing a building. The building must have a sturdy foundation or framework to survive. The geospatial community has settled on the term "Framework" to describe this foundation. Framework data is commonly used to describe seven themes of digital geographic data.

1. Geodetic control;
2. Orthoimagery;
3. Elevation;
4. Transportation;
5. Hydrography;
6. Governmental units; and
7. Cadastral information.

The National Digital Orthophoto program (see Box 2.2) illustrates how a Framework data theme can be completed.

There are commonalities between the USGS's proposed *National Map* content and Framework themes. The Geospatial One-Stop project (see Box 1.4), which relies heavily on existing structures within the FGDC, aims to accelerate the development of these themes, as recommended in the NRC report *The Data Foundation for the National Spatial Data Infrastructure* (NRC, 1995) (see Appendix C). Furthermore, Executive Order 12906 (see Appendix C) stipulates that FGDC is to consult with state, local, and tribal governments and submit a plan and schedule to OMB for completing the initial implementation of a national digital geospatial data "framework" by January 2000.[5]

---

[5]This has not yet happened (Milo Robinson, FGDC, personal communication, 2002).

The FGDC has endorsed the concept of Framework data (FGDC, 2002f).

The framework is a collaborative community based effort in which these commonly needed data themes are developed, maintained, and integrated by public and private organizations within a geographic area. Local, regional, state and federal government organizations and private companies see the framework as a way to share resources, improve communications, and increase efficiency.

With specific reference to the Framework data content the FGDC states that

[t]he framework represents "data you can trust"—the best available data for an area, certified, standardized, and described according to a common standard. It provides a foundation on which organizations can build by adding their own detail and compiling other data sets.

The potential carried by Geospatial One-Stop to accelerate development of the Framework layers offers a means of fueling advancement of USGS's vision. Lessons learned in implementing partnerships across all levels of government through Geospatial One-Stop, in addition to those learned by the FGDC over the last decade, will be invaluable to the USGS as it continues its mapping activities beyond the lifespan of Geospatial One-Stop. For example, the proposed "data-acquisition bazaar" may evolve into a national marketplace for data acquisition that promotes economically efficient allocation of financial resources at all levels of government (e.g., see comments of Scott Cameron, Appendix D). The USGS will need to work within the FGDC structure as it proceeds with partnerships and standards development relating to *The National Map*.

PROPOSED DATA CHARACTERISTICS

The USGS has laid out a series of goals for the characteristics of data layers within its *National Map* (see Chapter 1 and USGS [2001]). As with any such exercise, a number of challenges must be addressed to attain these goals. Issues, challenges, and questions raised by workshop participants and the committee on these data characteristics are summarized in Table 3.2. There was broad agreement among workshop participants and the committee that the characteristics outlined by the USGS for *The National Map* are valuable to the nation.

TABLE 3.2 Issues, Challenges, and Questions Relating to the Proposed
Characteristics of Data Layers in *The National Map*

Data Characteristic	Issues/Challenges/Questions
Current (seven days)	*Issues and Challenges:* This is the most challenging characteristic. It would also be a great asset. The challenges are largely institutional. Tracking the transactions would require extraordinary levels of collaboration. Although the seven-day schedule may be demanding,[a] there are many application domains that require real-time data.[b] *Questions:* • What are the triggers of change and updates? • Who reports a change? What is the role of volunteers? • What is the baseline and when is it updated? • How does currency vary for different themes? • How will the database support real-time applications (traffic, disasters)? • What constitutes a transaction? • How will users be notified of changes? • Can a geographic subscription service be provided (pushed to the subscriber)? • How are changes archived?
Seamless	*Issues and Challenges:* It is desirable for the USGS to move from a quadrangle-centric view of spatial data to one without artificial map sheet boundaries (i.e., accessible for arbitrarily defined study areas). Technical solutions already exist that enable spatial databases to function seamlessly. When dealing with large areas, the data volumes can be enormous, and bandwidth will be a real constraint. *Questions:* • Is TIGER modernization a good model to follow? • Can edge matching be handled automatically? • How to ensure that features are not duplicated or do not stop artificially? • How is seamlessness defined with respect to data source, scale, and time? Can a data layer compiled in a patchwork quilt model over time ever be truly seamless? • How to maintain consistent classification of features? • How will adjacent data from many sources be integrated?

Data Characteristic	Issues/Challenges/Questions
	▪ How will metadata be displayed for possible different sources across any given geographic expanse? ▪ How can the Geographic Names file be processed to remove tile or quadrangle dependencies?
Consistently classified	*Issues and Challenges:* Consistent classification standards are essential. The relevant FGDC standards must be followed and described in the metadata. Classification schemes should adhere to professionally adopted and disciplinary standards by theme. There will also be a need for translators, crosswalks, and support for alternative user-profile systems. In a Web environment a user profile could enable a user to develop a personalized or customized view of *The National Map* ("My National Map"). For example, some users may want to group land cover into simple categories such as "developed," whereas others may want "residential."
Variable resolution	*Issues and Challenges:* Variable resolution is a practical and cost-effective characteristic. There are several technical issues that need to be addressed and the resolution of any given data set for any area must be defined in the metadata. The USGS should approach resolution as a federal issue first (not local or state). FEMA suggests 1:12,000-scale (based on DOQ) for general mapping, but floodplains need higher-resolution data. The edge-matching problems caused by variable resolution will be severe and not always solvable. Additionally, there will be classification issues (e.g., land cover classification levels changing with resolution). Minimum mapping unit size will affect the classification, and problems related to classification of different-resolution source materials will need to be addressed. Lastly, there will be cartographic problems (e.g., appearance). *Questions:* ▪ What are the minimum scale requirements: 1:24,000? 1:12,000? Is the plan to begin with the 1:24,000 topographic series or will the USGS begin with remote sensing, where the problem of edge matching is not an issue? ▪ Are we limiting "resolution" to only raster data? Note that resolution and accuracy can affect attributes information and apply to raster and vector data. ▪ Is it desirable to have resolution vary by theme?

TABLE 3.2 Continued

Data Characteristic	Issues/Challenges / Questions
	▪ How to integrate multiple resolution data with variable positional accuracy?
Complete	*Issues and Challenges:* If *The National Map* is to be a trusted source, it must strive to be complete. That means every specifiable spatial feature should be included. The Spatial Data Transfer Standard (SDTS) includes a comprehensive list of spatial features that should be followed. *Questions:* ▪ What is the spatial extent of map coverage (i.e., does it cover Alaska, U.S. Territories, overseas military bases)? ▪ There are definitional problems as well as problems with minimum size: features, spatial extent, attribution, status, or state minimum mapping units. For example, what constitutes federal property? Would every post office be included?
Consistent and integrated	*Issues and Challenges:* The issues of consistency and integration are significant ones.[c] When users access the data, they need to have confidence that the data can be used in their applications with minimal manipulation. Themes that are logically connected (e.g., watershed boundaries, digital elevation models, streams) must be positionally correct to support analytical operations, and there should not be any cartographic displacement. Feature data as opposed to layers would allow one feature to be included in many contexts, greatly facilitating updates. *Question:* ▪ Are the layers logically integrated (e.g., drainage-enforced digital elevation models)?
Variable positional accuracy	*Issues and Challenges:* As with the notion of variable resolution, it is practical and cost effective to capture features from different-scale source material for urban and rural areas. Therefore, the positional accuracy of features will vary. The important issue is to report the reliability in the metadata. The FGDC standard for transportation features is a good model for other feature types.

Data Characteristic	Issues/Challenges / Questions
Consistent spatial reference	*Issues and Challenges:* The spatial reference system should meet standards for point positioning established by the SDTS. A national system is best served by a geographically referenced system based on latitude and longitude adjusted to a modern datum (ultimately, users must be able to select various map projections, and the latitude-longitude system is the best starting point). The national system should be NAD 83 with coordinates reported to seven decimal places. The vertical dimension should be referenced to NAVD 88. Most states require state plane coordinates; therefore there must be education and training about best practices in terms of projection changes and differences in U.S. survey and international feet. The USGS should play a leadership role in educating, providing software, and certifying accurate coordinate system, datum, and projection transformations.
Standardized	*Issues and Challenges:* The issue of standardization is paramount and is mandated by OMB Circular No. A-16 and other policies. Data elements must be endorsed by the FGDC.
Documented (FGDC metadata)	*Issues and Challenges:* FGDC and ISO (International Organization for Standardization) metadata standards should apply. To be trusted and valuable to the broadest community it should include feature-level metadata, which is particularly important for the temporal dimension, reporting accuracy, and tracking lineage.[a,b] *Question:* ▪ What is the most useful level of detail and who would provide the detail?
Temporally tracked and archived	*Issues and Challenges:* Tracking changes to features on the Earth's surface is a core scientific competency of the USGS, and this approach should be supported. This approach is the only cost-efficient way to support longitudinal studies, to do systematic archiving, and for rapid update. There are many practical problems related to how to implement and maintain a temporally accurate database. There must be a practical business model to support this, or it will not be reliable.

TABLE 3.2 Continued

Data Characteristic	Issues/Challenges / Questions
	Questions: ■ What should be the temporal interval for archiving and tracking (monthly, annually), and who captures, synchronizes, and posts the changes? ■ What type of archive methods will be used (retired features, snapshots, versioning)?

[a]See Appendix D, comments of Robert Marx.
[b]See Appendix D, comments of Donald Cooke.
[c]See Appendix D, comments of William Craig.
[d]See Appendix D, comments of Curt Sumner

ARCHIVING DATA

The suite of systems designed to ensure the reliability and surviv-ability of the content for *The National Map* and enhanced National Atlas will likely parallel those developed for paper maps and now applied to digital map archives. Past digital map data now represent on the order of petabytes of information, and the needs of the various archive systems must be taken into account at the outset for the long-term effectiveness of the contents. One concern is the question: Can the specific contents of *The National Map* and enhanced National Atlas be reconstructed effectively as they existed at a specific point in time? If so, not only will the survivability of the data be improved, but a new field of research and even business may form to understand and track the dynamics of geographic phenomena over time. For example, one private digital map company conducts considerable business by being able to recreate the ownership and land use of a single cadastral parcel or city block over long periods of time (Sanborn, 2002).

There is a federal role in data archiving in addition to the existing NSDI clearinghouse websites and potential private-sector roles. The Federal Depository Library program (FDLP) was established by Congress to ensure that the American public has access to federal government infor-mation. The FDLP involves the acquisition, format conversion, and distri-bution of depository materials and the coordination of Federal Depository Libraries in the 50 states, the District of Columbia, and U.S. territories. The mission of the FDLP is to disseminate information products from all three branches of the government to nearly 1,300 libraries nationwide. Libraries that have been designated as federal depositories maintain these information products, including maps and digital map data, as part of their existing collections and are responsible for assuring that the public has free access to the material (see comments of Ernest Baldwin, Appendix D, and GPO [2000]).

There is an established base, in a paper environment, for disseminating and archiving USGS and other spatial data through the FDLP. Following the congressional mandate in 1995 for GPO to change to an all-electronic FDLP, the minimum technical requirements for public access workstations in federal depository libraries (GPO, 2002a) required libraries to purchase equipment and develop the ability to provide access to the electronic data they were receiving through the FDLP. This required a significant investment by libraries to retain their depository status, and training has been provided at annual federal depository conferences in Washington, D.C.

The FDLP role could be expanded and extended by the USGS concept as proposed. Two potential new elements would be the need to interact with local, tribal, state, and private agencies that are data holders but whose data is searchable by public domain metadata; and the requirements for transaction-based updates. The latter, an alternative to versioning, will require research. Partnerships with FDLP libraries, GPO, and other agencies and groups (e.g., Digital Library Federation, National Science Foundation, and national supercomputer centers) could be sought by the USGS to develop appropriate archiving methods and sites. The ongoing FDLP Partnership program (GPO, 2002b) provides examples of archiving partnerships, albeit not for spatial data. In this program GPO facilitates partnerships with federal agencies from its network of FDLP libraries.

The archiving of spatial data is a continual challenge, and assuring the reconstruction of data at any given point in time, as well as ensuring the data's continuing viability through time, add to that challenge. The technology developed to meet this challenge will have far-reaching applicability, and the benefits of timely, accurate, and accessible data cannot be overstated.

VOLUNTEERS

Another component of the USGS proposal is the collection of data from users acting as volunteers. The contribution of data by private citizens (including school children on "GIS days"), although perhaps of value for publicizing and enhancing *The National Map*, best services the broader body of contributed uncertified data linked to the user through the NSDI similar to ESRI's Geography Network.

Volunteer programs are not alien to federal programs. For example, since 1994 the Environmental Protection Agency's Oceans and Coastal Protection Division has used volunteers for estuary monitoring (EPA, 2001). Citizen contribution of USGS certified data could meet with disapproval from professional surveyors, however (e.g., see comments of Curt Sumner, Appendix D). In an alternative approach the Delaware *National Map* pilot project (see Table 2.1) was able to use citizen data corrections effectively, with an e-mail feedback system to correct errors, and volunteers from the nonprofit United States Power Squadrons submit nautical chart corrections to the National Ocean Service (USPS, 2002). Such a system would be a good model for a volunteers program, whereas certification is beyond the scope of the federal government. The USGS will need to carefully review the role of volunteers. Any solution will need to be worked out with input from relevant professional organizations.

THE PUBLIC DOMAIN

The concept of the public domain is important in any discussion of the role of government information in promoting a ubiquitous information infrastructure. The USGS proposes in its vision document to place all data in the public domain (USGS, 2001). Edward Samuels defines the public domain in the U.S. setting as "works" divided into the following categories:

> [W]orks for which the term copyright has expired; works that are otherwise eligible for protection, but in which copyright has been forfeited because of a failure, either by design or by mistake, to comply with the copyright formalities in effect at the time; and works that are non-copyrightable because they are categorically excluded from federal copyright protection (Samuels, 1993).

Public domain map data have played a key role in stimulating innovation in the multibillion-dollar geographic information industry, one in which the United States leads the world. This role is remarkably consistent with the spirit of "public domain" in the 1884 National Academy of Sciences report, though the meaning has changed radically. Clearly, national goals and priorities should override those of individual agencies. It would be poor policy to hold back public spatial information for purposes of cost recovery or for private licensees, a lesson that was learned at great expense in past experiments with commercialization (Brown, 1997). The marketplace has been driven by low-cost public datasets that can be enhanced to serve a variety of end uses. The USGS must recognize the requirements of commercial data vendors if they are to promote value-added processes resulting from *The National Map* and enhanced National Atlas. For-profit and not-for-profit intermediaries play a significant role in a spatial data infrastructure by providing spatial data products and services that cannot be provided by government suppliers. For-profit enterprises add value to existing government information, such as improved detail, rearranged content, improved quality, or commercial applications. Currently such cooperation is only beginning to evolve. Commercial data providers are obvious partners in the USGS vision and have much to contribute.

Interorganizational partnerships, such as Cooperative Research and Development Agreements (CRADAs), between government agencies and for-profit and not-for-profit organizations require new transaction and valuation methods for exchanging information commodities. Public agencies should be careful in applying market approaches, and there are examples of failures for this reason (e.g., NCGIA, 1993). Basing the dissemination of public information on market concepts rather than on efficiency, effectiveness, fairness, and equity may reflect unfounded concepts. Also, with the presence of inherent network externalities in electronic markets, the successful exploitation of information products and services is fundamentally different from more conventional products and services. Recognizing these factors is critical if policy makers are to avoid pricing practices and accompanying regulatory mechanisms based on principles designed for a previous (industrial) economy. If a public policy objective of the USGS is to promote the wide availability, use, and downstream commercialization of *The National Map*, it is appropriate to consider reducing intellectual property and pricing impediments as an incentive for third parties to acquire and commercialize these data. By reducing such impediments information policy is crafted to parallel the national objectives for funding basic research, while stimulating the growth of emerging information industries (Lopez, 1998).

Assumptions about the value of public information in the public domain are set forth in OMB Circular No. A-130. The underlying principle for these assumptions is that government spatial data resources provide a fundamental framework that sustains a horizontally and vertically integrated spatial data infrastructure that is open, diverse, and innovative. Such a public infrastructure is critical to the development of a marketplace of ideas, products, and services.

A final consideration in the discussion of the public domain is data security. For example, local governments may want to integrate private utility data into local datasets for public safety or infrastructure design programs. Without assurances that their critical data would be publicly unavailable, full cooperation of local entities with the USGS program is unlikely. The USGS will need to review issues related to security of proprietary and other limited access data in conjunction with the public domain. While the coarser nature of the enhanced National Atlas scale will protect much of the content, other means of addressing security will need to be researched and clarified to all potential partners.

SUMMARY

The committee sees advantages in labeling the integrated, nationally consistent database (the "blanket") as the enhanced National Atlas, building from the success of an existing program. *The National Map* would be a "patchwork quilt" of data and metadata contributed from local, state, tribal, and private agencies. As a result of partnership arrangements coordinated by the USGS, these data would be submitted to the USGS for checking and some would gain official inclusion in the enhanced National Atlas. The partnerships would leverage existing data, reduce duplication of effort, and improve timeliness of updates.

Content for *The National Map* and the enhanced National Atlas should be thematically identical at their core but allow for variants in scale and specifications at the local level. The pertinent themes include the Framework and base cartographic layers and the thematic layers identified in OMB Circular No. A-16. These are already the subject of discussion and standards development within the FGDC, and there is much to be gained by using these specifications for the thematic content of *The National Map* and enhanced National Atlas. The data should be in the public domain and should be integrated into the Federal Depository Library program. Lastly, there may be value in incorporating direct input from the general public, especially in detecting errors and making updates.

4

Implementation of a National Map

INTRODUCTION

This chapter reviews the next steps proposed by the USGS in their vision of *The National Map* and outlines the implementation challenges, likely roles of partners, needed changes in USGS culture, and research needs. The chapter ends with a discussion of user requirements, since these will need to be met for the project to be successful.

NEXT STEPS FOR THE USGS

The USGS vision document (USGS, 2001) outlines a series of "next steps" in lieu of an implementation plan. These next steps are seen as dependent upon research and development. Short-term goals are stated to be attainable using existing technologies. Mid- and long-term goals are stated to be dependent upon research in applied topics within cartography, geographic information science, remote sensing, and information science.

The next steps foreseen by the USGS include a formal review (which includes this report), the alignment of USGS activities with the vision, and the forging of the relationships necessary to create the partnerships to shape the vision. The core of the implementation plan is to

   ▪ complete a five-year plan for the USGS Cooperative Topographic Mapping program, including setting data architecture, operations, goals, and timelines;
   ▪ develop a business model for implementation;
   ▪ identify and make necessary changes in business practices;
   ▪ identify workforce implications;
   ▪ investigate existing holdings and data partnerships for compatibility with the concept, and define data content;
   ▪ expand the state liaison activities by building pilots; and
   ▪ identify needs for legislative initiatives as implementation progresses.

   The absence of a thorough implementation plan can only be seen as an impediment to progress. At the USGS "Status of *The National Map*" website (USGS, 2002e), a graphic indicates that (1) a consistent user interface is to be developed by October 2002; (2) by February 2003 there are to be multiple Web map services and consistent theme descriptions; (3) by August 2003 a consistent symbology will be decided; and (4) by September 2003 a digital data extraction and graphic generation capability will exist. The vision document (USGS, 2001) states that "significant accomplishments must be programmed for the fiscal year 2002-06 timeframe"; *The National Map* will be accomplished in phases; and "the goal for full implementation of *The National Map* by 2010 is part of the vision."

   A more detailed vision is necessary for an effective implementation plan. Although the goals to be attained and the roles to be played in developing *The National Map* are discussed at length in USGS (2001), the specifics cannot be left to develop solely through pilot projects. Although pilot projects are an important first step for testing ideas and approaches, potential partners in the broader geospatial community need an understanding of the specifics of how they could be involved, the benefits to them, and the resources they will likely need to contribute if they are to buy into the concept. There are few barriers for the USGS in creating such an implementation plan, and indeed, progress may already have been made beyond the documents circulated as part of the visioning process. The discussion in this chapter includes issues relating to implementation that are developed as recommendations in Chapter 5. Foremost is the need to identify and highlight the major challenges the USGS will face in implementing the concept of *The National Map*.

CHALLENGES

The main challenges to implementation are organizational rather than technical. The USGS and FGDC have a history of developing standards for spatial data, and the USGS has decades of experience in digital cartographic data. By working with other federal agencies, perhaps in the existing FGDC working group structure (e.g., see comments of Patti Day, Appendix D) rather than through a separate federal advisory committee, USGS could develop both a design and an implementation plan, and should do so as rapidly as possible. Rapid development of an implementation plan is critical because it will increase the chance of leveraging existing data so that much can be accomplished in a short time.

The initial version of the enhanced National Atlas would be the existing digital map holdings of the USGS, enhanced with public domain data from elsewhere in the federal government, such as TIGER, the National Shoreline database, and the BLM's Geographic Coordinate database. The enhanced atlas can therefore be initiated by reorganizing existing holdings, and by bridging government departments, in the way envisioned in Geospatial One-Stop (see Box 1.4).

The initial version of *The National Map* would consist of immediately accessible, best available data from tribal, state, and local public domain sources (including ongoing pilot projects). A large number of U.S. cities are being overflown to collect high-resolution air photography that will become the basis of new federal and local datasets as part of the homeland security activities under way with NIMA (National Imagery and Mapping Agency) support. It would be counterproductive not to include the fruits of this project in *The National Map*.

Thus the initial challenge, that of specification of content and moving beyond pilot studies, can be overcome with budget support. A far more complex challenge is to create the new role for the USGS and its partners as described in the vision document under "next steps" as "[forging] relationships with organizations interested in *The National Map* vision." One aspect of this challenge is the large number and varied structure of local GIS systems (including a variety of procedures, data elements, and data dictionaries). The USGS will need to develop a wide array of partnership and implementation models that address the cultural and legal issues surrounding locally developed spatial data. Local entities may be governed by rules and statutes imposed by states or their own jurisdictions. This challenge is further complicated by a trend toward regionalization of local GIS systems through metropolitan planning agencies, interjurisdictional consortia or other voluntary associations that may have the status of local government. Ownership of spatial data may in some cases be difficult to determine or may require the approval of multiple entities to make local data available. It is unlikely that a single comprehensive partnering model will be effective in securing substantial local participation in *The National Map*.

ROLES

The USGS vision document considers that "partnerships are the key to the success of *The National Map*." Partners include other federal agencies, tribal, state, regional, and local governments, private industry, academia, libraries, and the general public. In each case the USGS has prior experience in building working relationships. The review stage of the USGS *National Map* vision has been open and available for each of these groups, primarily through the Internet but also through published outlets. There are many potential nontraditional partners who generate or use spatial data and should be involved. For example, the Centers for Disease Control and Prevention will play a critical role in bioterrorism prevention, and spatial elements of this challenge would be enhanced through data partnerships.

The committee agrees with the USGS that partnerships are key to success. Such a philosophy is embedded in past reviews of mapping activities (see Appendix C) and in the rules under which the USGS and other federal agencies operate. Given the obvious need for a national map (see Chapter 2), the question raised by the committee was why does such a map not already exist, and what has prevented it from existing? A review of the prior recommendations in Appendix C shows that the road to the concept of *The National Map* has been paved with good intentions.

The USGS should play the roles mandated in OMB Circular No. A-16 (see Box 2.1) with coordination through the FGDC. Central to the proposed USGS activities are the coordination of data acquisition, creation of effective incentives for local and private contributions to the NSDI through partnerships[1] and other means that assure timely and accurate maps, and coordination with other federal agencies. Since the revision of Circular No. A-16 in 2002 (OMB, 2002), the USGS is required to play the coordinating role that it now proposes in the vision document. Because this goal has been only partially fulfilled, how can the USGS

---

[1]See comments by Shoreh Elhami and Hugh Archer in Appendix D on the inclusion of all levels of government and some ideas for resources and arrangements for these partnerships.

increase its efforts to build *The National Map* and enhanced National Atlas through partnerships?

Processes and protocols will be needed to ensure that successive tiers of government or organizations can serve as providers of datasets within their jurisdictions. Each level of government would help the next, more local level. It will be vital to involve the local level in revision and maintenance if the seven-day update goal of the USGS is to be met.

GIS and mapping technologies are swiftly becoming core functions of government, particularly at the state and local levels. Many state and local governments have made GIS a part of their information technology environments, extending its use throughout their organizations as a valuable analysis and management tool. The movement of GIS to mainstream technology status (e.g., McGarigle, 2000) requires that organizations manage this technology the same way they manage other enterprise systems.

Many of these governments are becoming aware of the value of information architecture as a discipline that permits the rational and efficient development of enterprise-wide information systems. One of the greatest benefits of information architecture is the ability to plan for development of integrated information systems that allow information from disparate systems to be exchanged with relative ease. The USGS concept of *The National Map* depends in large measure on the ability of many organizations to exchange data. As governments move to adopt formal information technology architectures, it is essential that GIS and spatial data be identified as components of that architecture, and that information technology professionals be encouraged to develop architectures that will accommodate the requirements established for *The National Map*.

Federal programs that seek links with tribal, state, and local governments often have used distributed organizational structures to build these links. The USGS has a regional structure but also has a large number of field offices that the USGS already acknowledges could play this role. Many federal agencies have used states as the first tier of area integrators (e.g., see comments of Gene Trobia, Appendix D). An area integrator is a tier between the source of the finest spatial resolution data and the federal government. The USGS's regional structure is a tier above that of the state, though the current pilot projects have a stronger state orientation (see Table 2.1). The USGS vision document refers to area maintenance offices and field centers as two tiers that are envisioned for *The National Map*.

The task will be enormous for area integrators to evaluate existing data and integrate data from hundreds, perhaps thousands, of institutions to achieve the consistent seamless base map themes, and to integrate, certify, and then redistribute them. The USGS will need to consider how to ensure consistency and quality through training and cross-checking of data. Area integration is doable, however, as demonstrated by the USDA, which has started area integration of spatial data by county to support its National GIS Implementation plan (USDA, 2002).

The USGS should clearly identify a distributed organizational structure, using its existing regional tier, which allows effective area integration at the right level of spatial aggregation. In so doing, the bottom-up component of *The National Map* and enhanced National Atlas needs to be clearly recognized. Lessons from the pilot projects (see Table 2.1) show that Delaware is too small and Texas too big for effective area integration, in the committee's view. The USGS and its partners need to identify the most effective spatial distribution of integration, and how it relates to each partner's regions, recognizing that these will almost certainly overlap. Efficiencies and better coordination will be realized if USGS integration offices are managed in partnership with other state integration offices (such as state geographic information councils that currently perform interagency coordination functions) and other distributed federal offices and centers.

CULTURE CHANGE

The culture of historical paper map production is a counterforce to the type of cultural reinvention necessary for the USGS to change its role. To achieve its goals the USGS will need to phase out lithographic map printing, with the understanding that new user studies and research will investigate innovative alternative means of spatial data delivery to users. Hardcopy printing will likely need to be retained but improved. Regardless, the static quadrangle will need to be replaced by dynamic content and new print-on-demand technologies (see Chapter 3).

The greatest culture change, however, will be to reorient the USGS toward the proposed partnerships with a level of aggressiveness and willingness to cooperate that the USGS acknowledges to be unprecedented (see also comments of James Plasker in Appendix D). The USGS notes on page 16 of the vision document that organizational issues "will be among the most challenging ..." The "next steps" component of the vision document implies that this reorientation will require retraining; reallocation of workforce; new roles to be played, such as promotion and stewardship; new contacts with a multitude of agencies and volunteers; and a new suite of applied research projects.

RESEARCH NEEDS

Over the years, the USGS has played a role in cartographic research. Diminished funding and a revised role will require the agency to seek out further cooperative research and development agreements with the private sector and to work more closely with academic institutions and other organizations. The USGS has accomplished some research through its own Office of Research, but these capabilities will need to be rebuilt if the vision of *The National Map* is to be realized. The research needs are many (e.g., USGS, 2001; NRC, 2002). In the committee's opinion the following questions will need to be addressed in addition to those identified in Appendix 2 of the USGS vision document.

Feature extraction. Can computers detect, recognize, and then automatically represent in spatial databases features captured in high-resolution imagery? Can automated methods be devised to detect, measure, and update changes over time? Can methods be devised to anticipate and prioritize areas where changes are of higher frequency and magnitude?

Validation, quality control, and accuracy assessment. Can automated procedures be developed that process and evaluate spatial data without human intervention? Could these systems be used to apply the "seal of approval" concept to a spatial data custodianship role? Can quality control procedures be nested or even continuous in their approach, rather than simply "accept" or "reject"? How can spatial variation of error be modeled, how is it introduced during processing, and how might it be reduced (NRC, 2002)?

Watermarking and steganography. Are methods that control spatial data access sufficient to protect intellectual property and ownership? Can methods be devised and applied that allow proprietary and public data to coexist without theft or risk of security breaches?

Technology. What forthcoming technologies will affect the design and content of *The National Map*, and how can these changes be anticipated and planned?

Media. What media are suitable for storage of *The National Map*? What role will compression play (e.g., NRC, 2002)? How much redundancy (such as mirror sites) is necessary to ensure uninterrupted service?

Volunteer system. If members of the general public can submit data into *The National Map* (or at least the NSDI), what sorts of systems are necessary to validate and include these data in *The National Map* and atlas when these data meet standards? What portal is suitable for the reporting of such updates? Would engaging a wide range of volunteers help to build support for a *National Map*?

Large distributed database security and reliability. How can the components of *The National Map* be made secure and reliable?

Barrier-free access to information. How can we continue to expand the use and application of spatial information to users who may have physical limitations and assure compliance with Section 508 of the Americans with Disability Act (e.g., see comments of Ernest Baldwin and Yves Belzile, Appendix D)?

DETERMINING USER REQUIREMENTS

A key issue for any mapping solution that offers data products is an understanding of the "engines" that require the data products as "fuel" (Ronald Birk, NASA, personal communication, 2002). There is considerable debate about who will be the potential users of *The National Map* and the enhanced National Atlas. Depending on how the databases are designed and how the content is specified, these resources could potentially serve most of the spatial data needs of local, tribal, state, and private sector, as well as federal users.

The USGS will need to do a requirements analysis of existing and potential users to support its data quality goals. Understanding the customer is key to success (e.g., see comments of Hugh Bender, Appendix D). Data requirements vary throughout the nation and among user groups. Indeed, they vary for each user; for example, local, state, and tribal users will have different needs during emergencies than for more routine operations, and will need to know what federal data are available to supplement their existing databases. A user requirements analysis will be a challenging but necessary step for the USGS.

The USGS will also need to develop a business plan with milestones and metrics for judging accountability and performance of those involved in the *National Map* effort. For example, themes should be monitored for reduction or elimination of duplication, perhaps using similar reporting techniques to those in Geospatial One-Stop (see Box 1.4). Redundant data production and maintenance would be considered a "cost." Using road features as an example, a user requirements analysis would define the attributes and characteristics of a viable set of street centerlines and reveal the business rules to create road features that meet user expectations. In general, the business plan would be enhanced by inclusion of cost-benefit analyses or projected benefits. Such a practice is commonplace at the local level to justify funding requests (Alan Leidner, personal communication, 2002).

The USGS's *National Map* vision is too oriented toward the needs of the federal sector. Although broad acceptance, support, and cooperation within the federal government are essential for the success of the project, much of the concept is based on reconstituting the 1:24,000 scale topographic quadrangles. The content and scale of those maps are useful at the scale of many federal activities and applications, however the 1:24,000 scale is too small for many tribal, state, and local applications. New data at this scale offer no incentive for participation by these levels of governments. To succeed, *The National Map* will need to be multiresolution, with data maintenance at the appropriate levels of government. Workshop participant William Craig suggested building on the NRC evaluation of FGDC partnership programs (NRC, 2001) that outlined responsibilities for data collection and maintenance (see Table 4.1). Craig suggests that

> state and local government need to see some benefits to themselves for participating in this effort. The honor of recognition of their efforts will be appreciated, but their additional effort in cooperating will be substantial, including modifying existing work to fit new standards, even if the result is a better product. I see two potential benefits that would draw them in: returning PLSS corners to the vision of *The National Map* and cost-sharing on various projects of mutual interest, including orthophotography and high resolution digital elevation. USGS and its federal partners may feel unable to promise such matches, but state and local governments would be excited at the prospect of getting that assistance and willing to take the case to their members of Congress for support.

The USGS vision is also too USGS-centric. Partnerships are needed with other federal agencies, and the FGDC subcommittees may provide a forum for understanding federal requirements in addition to those of other potential users. NOAA's National Ocean Service (see comments from Anne Hale Miglarese, Appendix D) and the Bureau of the Census (Robert Marx, Bureau of the Census, personal communication, 2002) have endorsed the USGS vision of *The National Map*. Census has demonstrated that it can ingest data from local governments (e.g., see Box 2.3), and its approach provides a good model for what will be a critical factor in the ultimate success of *The National Map* and enhanced National Atlas: participation at the local level.

TABLE 4.1 Possible Responsibilities for Data Layers in an Integrated *National Map* and Enhanced National Atlas Setting

Theme	Possible Responsibilities		
	Federal	State	Local
Digital orthoimagery (scale dependent)	Primary[a] at coarse resolution	Supplementary	Primary at fine resolution
Elevation	Primary *at coarse resolution*	Supplementary for roads and state projects	Primary *at fine resolution*
Bathymetry	Primary for offshore	Supplementary for lakes and reservoirs	Supplementary for ponds
Hydrography	Primary	Supplementary	Supplementary
Transportation	Supplementary	Primary for highways	Primary *for streets*
Government units	Primary for states and international	Primary for state-owned parcels and counties	Primary for municipalities
Boundaries of public lands	*Primary*	*Supplementary*	*Supplementary*
Structures	*Supplementary*	*Supplementary*	*Primary*
Geographic names	*Primary for cultural features*	*Supplementary*	*Primary for street names*
Land cover *and land use*	Primary for land cover	Supplementary for both	Primary for land use

TABLE 4.1 Continued

| Theme | Possible Responsibilities | | |
	Federal	State	Local
Cadastral information	Primary for PLSS, leases, and easements on public lands	Supplementary	Primary
Geodetic control	Primary	Supplementary	Supplementary

SOURCE: Adapted from NRC (2001) by workshop participant William Craig, with minor additions from the study committee on geographic names and elevation. *Italicized* text denotes additions to the original table in NRC (2001). [a]"Primary": supersedes other sources for most themes but orthoimagery and elevation can have multiple representations.

SUMMARY

The USGS's vision document proposes completion of *The National Map* by 2010. To achieve this, USGS will need to work on product definition, specifications, and an implementation plan. Too many of the vision's central questions have been postponed as decisions for implementation. The USGS should, as it indicates it will, begin work immediately on producing such a plan. Nevertheless, the opportunities, data, technology, and will to accomplish the building of the enhanced National Atlas already exist.

More difficult will be adapting the USGS into a new role, that of data integrator, guarantor, certifier, and custodian of the integrated base map data. This USGS role is already mandated by OMB Circular No. A-16, and the committee estimates that if the 1990 version of A-16 had been fully implemented and funded by OMB, the nation would now have nationally consistent, integrated data themes at 1:24,000 (and perhaps 1:12,000). The cultural change required at USGS will be extensive and will involve moving completely away from paper products toward a new distributed model in which data are simultaneously produced, integrated, updated, and redistributed. In the past these tasks were separated by years.

The USGS has done a good job of assessing its research needs, although there are additional needs. The USGS should rebuild its research capacity and seek partnerships with other federal agencies, industry, and with academia to fulfill these research needs. The majority of the needs are technical and applied. Nevertheless, for long-term survival *The National Map* and enhanced National Atlas need a research agenda that focuses on technical challenges and the new options that these databases create.

Many of the data development, data maintenance, and integration tasks currently believed to be USGS responsibilities may best be done in collaboration with partners. A user requirements assessment, perhaps conducted through the FDGC structure, will increase project participation and buy-in and increase the value of the data holdings. An effective implementation plan will recognize that needs vary at the local tier, and that the focus should not be solely federal-down but also local-up.

5

Conclusions and Recommendations

INTRODUCTION

The chapter begins with the committee's overarching conclusions. The remainder of the chapter contains a series of recommendations that could, when embraced and implemented, ensure that the USGS enters the twenty-first century with a sound national mapping strategy. The recommendations are sequenced to highlight partnerships as precursors to further steps, reflecting the report's emphasis on partnerships for success of *The National Map*.

OVERARCHING CONCLUSIONS

The workshop participants and the committee recognize that the *National Map* vision of the USGS is ambitious, challenging, and worthwhile. Nevertheless, there is also a uniform sense that the project is not well defined and needs a thorough definition. Technically the project may be feasible; organizationally it will require a significant investment in restructuring and rethinking the systems that have changed little over the last two decades.

In this report the Committee on the U.S. Geological Survey Concept of *The National Map* discussed the similarities of *The National Map* and the existing National Atlas. The latter is described by USGS as being a component of *The National Map* (USGS, 2001). In large part the National Atlas has been built using coordination and partnerships, using a national standard to develop nationally consistent small-scale databases from larger-scale data. Data themes are owned and maintained by different federal agencies and updates are provided to the USGS for inclusion in the National Atlas. The same should be true of *The National Map*, though at larger scales and with more partners.

The USGS concept of *The National Map* has two principal components, each dependent on the other. We have used a blanket and patchwork quilt metaphor in explaining these two components. The blanket, which we have termed the enhanced National Atlas (to extend the existing program), is a consistent national digital map coverage maintained at one or more scales. This blanket coverage that includes Framework layers would be built from public domain data and broadly disseminated following the philosophy in OMB Circular No. A-16. The second component, the patchwork quilt *National Map*, would be the result of contributed imagery and maps from local, state, and tribal governments, and from private and nonprofit organizations, contributed as part of a sweeping collaborative effort. This quilt would consist of patches of larger-scale data adhering to national standards but with varied resolutions and filled with smaller-scale data from the enhanced National Atlas when no other source exists. Some of the data will be public, some proprietary with publicly accessible metadata.

The USGS would serve as the integrator for all map contributions, assembling and merging data, and certifying and issuing a "seal of approval" to data included in *The National Map* or as an update in the nationally consistent enhanced National Atlas. As the Delaware pilot has shown, the USGS goal of seven-day updates could be attainable using the enhanced National Atlas/*National Map* approach. Such a dynamic *National Map* will need to support multiple scales, resolutions, classifications, and feature types provided by *National Map* partners. It will also require extraordinary coordination.

Implementation of the enhanced National Atlas could be attainable in stages. Larger-scale Framework data at the 1:12,000 or 1:24,000 scale such as geodetic control, digital orthoimagery, Public Land Survey System data, and public ownership boundaries could become part of the enhanced National Atlas in the near term. Other data types, such as hydrography and transportation, may not be completed for several years, since they require significant integration with other data types at the 1:12,000 or 1:24,000 scale.

Implementation of the patchwork quilt component, *The National Map*, is insufficiently specified in the USGS's vision document to estimate a timeframe, and a cultural change at the USGS will be required if the project is to be genuinely national in scope. Much of the existing data at the local scale has not been developed using national standards and will require revision or recollection to be integrated with other datasets. However, new larger scale projects, such as the 133 Cities project (USGS, 2002f) in support of homeland security, can be completed using national standards and would be an early addition to *The National Map*. The USGS can also learn from organizational and technical challenges that have been tackled during implementation of the 133 Cities project.

THE ROLES OF THE USGS IN PARTNERSHIPS

1. Conclusion: *The National Map* concept is currently loosely defined. An explicit implementation strategy and definition of data characteristics is absent. The USGS vision document adds to the mix of already complex programs and terminologies, and reads as a USGS-specific document rather than a concept document for a compelling new national program that reaches far beyond a single federal agency. There is little new in the vision document that has not already been written or discussed as part of NSDI and the Framework program, and beyond what is already mandated for USGS by the recently revised OMB Circular No. A-16. Furthermore, there is at least a 10-year history of recommendations directed mostly at the USGS to build partnerships and Framework data, and yet the mapping mission is still at the conceptual stage. Some of the earlier ideas, if they had been implemented before today, could have led to completion of *The National Map*, as outlined.

An effective implementation plan should synthesize federal business requirements for geographic data, including (1) the role the data play in accomplishing agency missions and programs, and (2) the level of resolution, accuracy, completeness, and update cycle for each theme to meet these requirements. The plan should strike a balance between meeting all needs and what the federal agencies can reasonably fund, perhaps as partners in joint budget initiatives. It is critical to the success of the USGS proposal that the plan include methods and incentives, financial when necessary, for state and local government participation in data development in support of their missions, particularly with respect to public safety and emergency services involving police, fire and rescue, and 911 personnel (including methods for controlling data access where appropriate). The experiences of other federal agencies working in partnership with other levels of government will be a valuable resource to the USGS as the implementation plan is constructed.

In light of recent natural and human-induced disasters and the heightened interest in homeland security, there is a need to move quickly to an implementation plan. The nation has a vested interest in ensuring rapid implementation of a nationally integrated spatial database to meet national needs, including national security, environmental protection, and land stewardship. The benefits include not only more efficient use of natural resources, prevention of loss of life and property, and reduction of duplication and waste but also economic benefits to the geospatial business community. Participation in and use and maintenance of this database will require financial commitments from all partners to initiate and maintain the effort. The multiplier effect of these investments could be significant and could promote continued U.S. leadership in this field.

Recommendation: The USGS should move expeditiously to develop an implementation strategy for its *National Map* concept in collaboration with USGS's many partners. The strategy should be clear on the needs, roles, incentives, and projected costs for all partners, on goals, milestones, and responsibilities, and on the USGS role with respect to FGDC activities, Geospatial One-Stop, and other initiatives to build out the National Spatial Data Infrastructure. The draft implementation plan should be circulated to all FGDC members and partners for comment.

2. **Conclusion:** *The National Map* as presently conceived is a large, ambitious project. Its success depends upon a number of factors that are beyond the control of the USGS. As a general approach to the project, the USGS should continue to build from a more modest, step-wise series of activities that are readily attainable, such as its pilot projects. The committee sees the development of integrated base geographic information for the nation as a cultural and institutional challenge more than a scientific or technical one. Tackling this challenge will require (1) the USGS Geography Discipline to be proactive in developing relationships at all levels of government, (2) significant engagement by USGS leadership, and (3) that the USGS critically examine its philosophy, structure, and processes. This new role is distinct from and builds upon the USGS's existing coordination role. The coordination role remains necessary, particularly in the areas of standards development and quality assurance, but a key question the USGS must ask its partners at every government level is how can the USGS assist them, and are these partners willing to provide resources to support the resulting identified needs and demands?

Recommendation: The USGS should make a priority of building the necessary partnerships for an integrated spatial database, while continuing to use small steps and pilot studies to gain experience in revision, integration, and updating procedures and partnerships. The pilot studies should be seen not only as technical but also as organizational and management prototypes. The USGS should place more Geography Discipline emphasis on building these partnerships to assemble Framework data through collaborative programs.

3. Conclusion: OMB Circular No. A-16 defines and assigns a leadership role to the FGDC, and assigns authority to the USGS for specific data themes. The FGDC is a focal point for coordination of federal, state, and local geospatial activities, and the *National Map* and enhanced National Atlas project will lead to an increasingly central role for the FGDC. OMB Circular No. A-16 assigns the chair of the FGDC Steering Committee to the secretary of the Department of the Interior and the vice-chair to the deputy director for management of the OMB, but allows delegation of this authority. The monthly FGDC Coordination Group meeting is chaired by the FGDC staff director. Because membership of the FGDC Steering Committee is at the departmental level, neither the director of the USGS nor the associate director of geography is currently involved as members of the FGDC.

Recommendation: The USGS leadership should increase its participation in FGDC processes to nurture the partnerships needed to accomplish its vision. The director of the USGS should participate in the Steering Committee meetings, and the associate director of geography should participate in the monthly Coordination Group meetings. The USGS should encourage the Coordination Group to include the executive program managers of federal mapping and geographic data-collection programs and the Steering Committee meetings to include the heads of the agencies that oversee these programs.

ENSURING PARTICIPATION IN AND WIDESPREAD USE AND MAINTENANCE OF *THE NATIONAL MAP*

4. Conclusion: To be in compliance with OMB Circular No. A-130 it is essential that federally produced data remain in the public domain. Copyright law states that copyright protection is not available for any work of the U.S. Government; "work" being described as that "prepared by an officer or employee of the [government] as part of that person's official duties." (U.S. Copyright Office, 2001). In addition to fostering broad usage of *The National Map* and enhanced National Atlas, data in the public domain will stimulate the economy by offering data to the commercial sector for enhancement and value-adding opportunities. The patchwork quilt model of *The National Map* would contain metadata for commercially and locally produced proprietary data as well as data in the public domain. In the blanket model for the enhanced National Atlas all data would be in the public domain.

Recommendation: The USGS in coordination and partnership with map-producing agencies named in OMB Circular No. A-16 and other levels of government and the private sector should develop an enhanced National Atlas with larger-scale map coverage. This should be freely available on the Internet and in libraries in the Federal Depository Library program (FDLP). State, local, and privately produced map data that are identified through metadata files within *The National Map* may have access restrictions that would require protocols or a fee. The metadata for this restricted data should, however, be in the public domain and freely available.

5. Conclusion: The maps produced by the USGS over the last 125 years are a significant contribution to scientific knowledge and research. Maps produced in the future will also need to be captured and retained. In one archival approach all changes would have metadata and be archived so that a map could be re-created for any point in time. Alternatively datasets could be archived so that changes over time could be tracked. In either case migrating data will be necessary to maintain accessibility and readability as new technologies evolve.

Libraries across the country are repositories for USGS paper maps. Map libraries, among them many of the 53 regional depositories of the FDLP, have maintained complete runs of all USGS maps from the first issues. Libraries and the data centers that have become commonplace therein could assist in preserving and providing access to digital map data. There is precedent for partnerships for archival preservation of digital data among the libraries in the FDLP, and each of these partnerships has been negotiated through the U.S. Government Printing Office.

Recommendation: Partnerships with FDLP institutions should be explored for digital archiving and archival maintenance of *The National Map* and enhanced National Atlas through cooperation with the U.S. Government Printing Office. The archival methodology should be the most efficient that technology allows, and a plan should be devised and implemented for continuity of the archive.

6. Conclusion: Digital spatial data are becoming a major focus of broader information management roles and technologies. Failure to recognize this trend could unnecessarily isolate mapping efforts from more general information management. *The National Map* and enhanced National Atlas will become critical components in the management, use, and dissemination of almost all federal government data. Better links should be developed between geographic information management specific to *National Map* and enhanced National Atlas databases and information technology in the broader sense.

Recommendation: Federal and state chief information officer councils and associations should be invited to participate in FGDC meetings and other strategic planning meetings for building out *The National Map* and enhanced National Atlas. The USGS and partners should work toward integrating geographic information into information technology enterprise architecture being developed and advanced by federal and state information technology architecture committees.

DATA CHARACTERISTICS AND METHODS

7. Conclusion: The committee endorses the USGS's plan for a nationally consistent set of base map data that includes pointers to multiple-scale map and image data, and that is sufficiently flexible to be subdivided into geographical units of direct interest to the users, such as congressional districts, counties, and watersheds. To build an effective database, partnerships are essential, with agencies responsible for such Framework themes as geodetic control, cadastre (not including private land ownership), and national coastline data. One of the functions of federal, state, local, and private partnerships in the USGS's plan will be to drive updates of an enhanced National Atlas through *The National Map*. Although ambitious, a seven-day turnaround may be achievable for certain data theme updates to be included in the enhanced National Atlas. Volunteer input could be acceptable if the users receive formal training or certification (perhaps by professional associations). A model of direct user participation is the Delaware *National Map* prototype, where such map errors as incorrect geographic names can be reported at a website or by e-mail. A seven-day turnaround is already possible using this system. Procedures set up in Delaware and experience from that pilot project could guide the USGS in its exploration of data update options.

Recommendation: Two synergistic organizational structures are needed for the USGS's contribution to building the National Spatial Data Infrastructure. The first is an enhancement of the existing National Atlas and includes Framework data (some of which already exists and will require partnerships with NOAA and BLM in particular). The data in the atlas should be public domain, at such a consistent scale as 1:12,000 or 1:24,000, and could be served through many existing and new gateway public and private Internet sites. The second structure, called *The National Map*, would serve users needing integrated larger-scale data, drive updates to the enhanced National Atlas, and implement many of the ideas that the USGS has proposed: seamlessness, voluntary contribution, a mix of public domain and private data, shared metadata, and nonuniform scale.

8. Conclusion: Successful implementation of an enhanced National Atlas and *The National Map* requires directed research aimed at problems that will be specific to the new approach. Much research can be accomplished by federal agencies, such as the National Imagery and Mapping Agency and the national laboratories (e.g., those of the Department of Energy and the Environmental Protection Agency). Others will need to be addressed through broad participation of such agencies as the National Science Foundation, and by nongovernment organizations. The USGS will need to rebuild its research capacity to conduct directed research on cartography and geographical science around the needs of *The National Map* and enhanced National Atlas.

Recommendation: The technical issues needing future research (in addition to those identified in USGS [2001]) include feature extraction, validation, quality control and accuracy assessment, watermarking and steganography, impact of new technologies, storage media, and database security and reliability. The USGS should investigate how this research can be carried out, by whom, and at what cost.

9. Conclusion: The methods and technologies for digital mapping, maintenance, and data access described in the USGS vision document show how the USGS will move to exclusively digital production of its paper map series and to partnerships as the primary production approach. Paper maps and in-house map production are no longer central to the mapping mission of the USGS. A plan is needed to address these shifting priorities and the costs and benefits of new technologies for delivering the same geospatial user services currently provided in paper form. Such a plan must address the conversion of staff, leadership, and culture so that the USGS moves from a federal agency that creates and maintains paper maps to one that (1) coordinates with other federal, state, local, and tribal agencies and with the private sector to create integrated, nationally consistent geographic data themes; (2) maintains custodianship of national spatial data; and (3) supplies geographic science and technical resources and assistance to partners at all government levels.

Recommendation: A component of the USGS's implementation plan should address the phasing out of updates and printing of paper 1:24,000 topographic maps. The digital integrated base map themes that will replace the paper maps should be available and accessible on the Internet for downloading and printing on demand through such portals as Geospatial One-Stop using Web mapping services technology. USGS should increase the use of the private sector in this endeavor and in providing technical support to partners.

References

Bhambani, D. 2002. Interior moves forward with Geospatial One-Stop. Government Computer News. Available at <http://www.gcn.com/ cgibin/udt/im.display.printable?client.id=gcndaily2&story.id=19801 >. Accessed October 14, 2002.

Brown, G. 1997. Remarks by Congressman George E. Brown at the Third Congress of the North American Remote Sensing Industries Association. Available at <http://www.house.gov/science_democrats/ speeches/remsense.htm>. Accessed November 20, 2002.

Bureau of the Census. 2002a. American Community Survey. Available at: http://www.census.gov/acs/www/. Accessed December 30, 2002.

Bureau of the Census. 2002b. Summary File 1: 2000 Census of Population and Housing, Technical Documentation Issues. Available at <http://www.census.gov/prod/cen2000/doc/sf1.pdf>. Accessed November 20, 2002.

Bureau of the Census. 2002c. Miscellaneous News Releases. Available at <http://www.census.gov/Press-Release/www/misc.html>. Accessed November 20, 2002.

Bureau of the Census. 2003. Subcommittee on Cultural and Demographic Data. Available at < http://www.census.gov/geo/www/standards/ scdd/>. Accessed January 29, 2003.

Cameron, S. 2002. Geospatial One-Stop: Update for Federal Geographic Data Committee. PowerPoint Presentation to FGDC on October 9, 2002. Washington, D.C.

EPA (Environmental Protection Agency). 2001. Volunteer Estuary Monitoring Program: Wave of the Future. Available at <http://www.epa.gov/ owow/estuaries/volmon.htm>. Accessed December 13, 2002.

Executive Order 12906. 1994. Coordinating Geographic Data Acquisition and Access: The National Spatial Data Infrastructure. Federal Register 59(71): 17671-17674.

FGDC (Federal Geographic Data Committee). 1997a. Framework Introduction and Guide. Available at < http://www.fgdc.gov/ framework/frameworkintroguide>. Accessed November 7, 2002.

FGDC. 1997b. A Strategy for the NSDI. Available at < http://www.fgdc.gov/nsdi/strategy/strategy.html>. Accessed November 7, 2002.

FGDC. 1999a. FGDC Coordinating Group Retreat Summary: Shepherdstown, WV. January 19-20, 1999. Available at <http://www.fgdc.gov/fgdc/docs/retreatsum/results_of_retreat.html>. Accessed November 21, 2002.

FGDC 1999b. Minutes of FGDC Steering Committee/Stakeholders Meeting. February 24, 1999. Available at <http://www.fgdc.gov/fgdc/steer/steer022499.html>. Accessed November 21, 2002.

FGDC 1999c. 1999 National GeoData Forum: Making Livable Communities a Reality. Forum Threads. Available at <http://www.fgdc.gov/99Forum/threads.html>. Accessed November 21, 2002.

FGDC. 2000. Improving Federal Agency Geospatial Data Coordination. Available at <http://www.fgdc.gov/fgdc/dst511.doc>. Accessed November 7, 2002.

FGDC (Federal Geographic Data Committee). 2002a. Federal Geographic Data Committee. Available at <http://www.fgdc.gov>. Accessed December 30, 2002.

FGDC. 2002b. Geospatial Information One-Stop Fact Sheet. Available at < http://www.fgdc.gov/geo-one-stop/docs/factsheet.pdf >. Accessed October 16, 2002.

FGDC. 2002c. Geospatial One-Stop: Best Practices White Paper. Available at <http://www.fgdc.gov/geo-one-stop/index.html>. Accessed October 16, 2002.

FGDC. 2002d. Federal Geographic Data Committee (FGDC) organization. Available at <http://www.fgdc.gov/fgdc/ fgdcmap.html>. Accessed November 15, 2002.

FGDC. 2002e. Subcommittee on Base Cartographic Data. Available at <http://www.fgdc.gov/sbcd/sbcd.html>. Accessed November 8, 2002.

FGDC. 2002f. Framework. Available at <http://www.fgdc.gov/ framework/ framework.html>. Accessed November 8, 2002.

GAO (U.S. General Accounting Office). 1982. Duplicative Federal Computer-Mapping Programs: A Growing Problem. GAO/RCED-83-19. Gaithersburg, Md.: GAO.

GPO (Government Printing Office). 2000. Depository Library Public Service Guidelines for Government Information in Electronic Formats. Available at <http://www.access.gpo.gov/su_docs/fdlp/ mgt/ pseguide.html>. Accessed December 30, 2002.

GPO. 2002a. 2002 Minimum Technical Requirements for Public Access Workstations in Federal Depository Libraries. Available at <http:// www.access.gpo.gov/su_docs/fdlp/computers/mtr.html>. Accessed November 22, 2002.

GPO. 2002b. FDLP Partnerships. Available at <http://www.access.gpo.gov/ su_docs/fdlp/partners/index.html>. Accessed December 30, 2002.

Keller, B., and G. Kreizman. 2002. To the Rescue: GIS in New York City on September 11. Research Note: Gartner Research. March 11, 2002. Available at <http://www.llgis.org/pages/news/docs/ GIS_tothe_Rescue.pdf>. Accessed November 6, 2002.

Lemen, R. 1999. The Evolution of Topographic Mapping in the U.S. Geological Survey's National Mapping Program: U.S. Geological Survey Open-File Report 99-386. Available at <http://pubs.usgs.gov/ of/of99-386/lemen.html>. Accessed November 20, 2002.

Lopez, X. 1998. The Dissemination of Spatial Data: A North American European Comparative Study on the Impact of Government Information Policy. Norwood, N.J.: Ablex Publishing.

McGarigle, W. 2000. Interoperability and convergence push GIS into information technology infrastructure. Government Technology. Available at < http://www.govtech.net/magazine/gt/2000/aug/ ProductFocus/ ProductFocus.phtml>. Accessed December 13, 2002.

NAPA (National Academy of Public Administration). 1998. Geographic Information for the 21st Century: Building a Strategy for the Nation. Washington, D.C.: NAPA.

NAS (National Academy of Sciences). 1884. Proceedings 1(2):141-146. Washington, D.C.: Judd & Detweiler Printers.

NCGIA (National Center for Geographic Information and Analysis). 1993. GIS, Cartography, and the Information Society: An Annotated Bibliography. W. Dowdy, comp. Available at <http:// www.ncgia.ucsb.edu/Publications/Tech_Reports/93/93-12.PDF>. Accessed November 20, 2002.

NIMA (National Imagery and Mapping Agency). 1997. Geospatial Information Infrastructure Master Plan. Available at < http://www.amso.army.mil/terrain/library/gii-mp/>. Accessed November 20, 2002.

NOAA (National Oceanic and Atmospheric Administration). 2002. Federal Geographic Data Committee Marine and Coastal Spatial Data Subcommittee. Available at < http://www.csc.noaa.gov/fgdc_bsc/overview/bathy.htm>. Accessed January 29, 2003.

NRC (National Research Council). 1980. The Need for a Multi-purpose Cadastre. Washington, D.C.: National Academy Press.

NRC. 1981. Federal Surveying and Mapping: An Organizational Review. Washington, D.C.: National Academy Press.

NRC. 1982. Modernization of the Public Land Survey System. Washington, D.C.: National Academy Press.

NRC. 1990. Spatial Data Needs: The Future of the National Mapping Program. Washington, D.C.: National Academy Press.

NRC. 1993. Toward a Coordinated Spatial Data Infrastructure. Washington, D.C.: National Academy Press.

NRC. 1994. Promoting the National Spatial Data Infrastructure Through Partnerships. Washington, D.C.: National Academy Press.

NRC. 1995. A Data Foundation for the National Spatial Data Infrastructure. Washington, D.C.: National Academy Press.

NRC. 1997. The Future of Spatial Data and Society. Washington, D.C.: National Academy Press.

NRC. 2001. National Spatial Data Infrastructure Partnership Programs: Rethinking the Focus. Washington, D.C.: National Academy Press.

NRC. 2002. Research Opportunities in Geography at the U.S. Geological Survey. Washington, D.C.: National Academies Press.

NRCan (Natural Resources Canada). 2002. An Evaluation of Economic and Social Impacts of NTS Topographic Maps. Available at <http://www2.nrcan.gc.ca/dmo/aeb/English/ReportDetail.asp?x=164&type=rpt>. Accessed November 21, 2002.

OMB (Office of Management and Budget). 1990. Circular A-16 Revised. Available at <http://www.whitehouse.gov/omb/circulars/a016/a016.html >. Accessed November 8, 2002.

OMB. 2000. Geospatial Information Initiative: IT Roundtable. Available at <http://www.fgdc.gov/fgdc/coorwg/2000/omb7_7summary.doc>. Accessed November 21, 2002.

OMB. 2002. Circular No. A-16 Revised. Available at <http://www.whitehouse.gov/omb/circulars/a016/a016_rev.html>. Accessed November 8, 2002.

Renewable Natural Resources Foundation. 1996. Congress on Applications of GIS to the Sustainability of Renewable Natural Resources. Renewable Resources Journal 14(3). Bethesda, Md: RNRF.

Samuels, E. 1993. The Public Domain in Copyright Law. Journal of the Copyright Society of the USA 41 (winter): 137-182.

Sanborn. 2002. Traditional Mapping. Available at <http://www.sanborn.com/services/traditional/traditional.htm>. Accessed November 20, 2002.

Thompson, M. 1988. Maps for America, 3d ed. Washington, D.C.: U.S. Government Printing Office.

U.S. Copyright Office. 2001. Copyright Law of the United States of America. Available at <http://www.copyright.gov/title17/>. Accessed November 20, 2002.

USDA (U.S. Department of Agriculture). 2002. Geospatial Data Gateway. Available at <http://lighthouse.nrcs.usda.gov/gateway/>. Accessed November 8, 2002.

USGS. 2001. *The National Map*: Topographic Mapping for the 21st Century. Reston, Va.: USGS. Available at <http://nationalmap.usgs.gov/report/national_map_report_final.pdf>. Accessed October 30, 2002.

USGS. 2002a. Office of Budget, Analysis of Activity. Available at <http://www.usgs.gov/budget/2002_Justification/01actanalysis.html>. Accessed November 20, 2002.

USGS. 2002b. USGS Mapping Partnership Program. Available at <http://mapping.usgs.gov/www/partners>. Accessed November 6, 2002.

USGS. 2002c. The National Biological Information Infrastructure. Available at <http://www.nbii.gov/about/>. Accessed December 18, 2002.

USGS. 2002d. USGS Fact Sheet 062-02 (June 2002): *The National Map* Pilot Projects. Available at < http://mac.usgs.gov/mac/isb/pubs/factsheets/ fs06202.html>. Accessed October 15, 2002.

USGS. 2002e. Status of *The National Map*. Available at <http://nationalmap.usgs.gov/nmstatus.html>. Accessed November 8, 2002.

USGS. 2002f. Homeland Security and *The National Map*. Available at <http://mac.usgs.gov/mac/isb/pubs/factsheets/fs06102.html>. Accessed November 15, 2002.

U.S. National Performance Review. 1993. Creating a Government that Works Better and Costs Less: Report of the National Performance Review. New York, N.Y.: Times Books.

USPS (United States Power Squadrons). 2002. Homepage. Available at <http:// www.usps.org/newpublic1/guesthome.htm>. Accessed December 13, 2002.

Appendixes

Appendix A

Biographical Sketches of Committee Members

Keith C. Clarke, *chair,* is a research cartographer and professor. He holds a B.A. degree with honors from Middlesex Polytechnic, London, England, and M.A. and Ph. D. degrees from the University of Michigan, specializing in analytical cartography. Dr. Clarke's most recent research has been on environmental simulation modeling, modeling urban growth using cellular automata, terrain mapping and analysis, and the history of the CORONA remote-sensing program. Dr. Clarke is the former North American editor of the *International Journal of Geographical Information Systems,* and is series editor for the Prentice Hall Series in Geographic Information Science. In 1990 and 1991 Dr. Clarke was a NASA/American Society for Engineering Education Fellow at Stanford University, and in 1992 served as science advisor to the Office of Research, National Mapping Division of the U.S. Geological Survey in Reston, Virginia. He is the Santa Barbara director of the National Center for Geographic Information and Analysis and chairs the Geography Department at University of California, Santa Barbara. He served as president of the Cartographic and Geographic Information Society for 2000-2001, and currently chairs the American Congress on Surveying and Mapping's Communications Committee.

Michael R. Armstrong is chief information officer of the City of Des Moines in Iowa. He is responsible for all voice and data systems in the jurisdiction. Since September 1997 he has led the development of the city's technical environment, including implementation of permitting and licensing, and enterprise GIS systems. A frequent contributor to national journals and speaker at regional and national meetings, Mr. Armstrong is a member of the Urban and Regional Information Systems Association, the Center for Digital Government, and the Metropolitan Information Exchange. He is technical advisor to the Des Moines/Cherkasy (Ukraine) development partnership and chair of Public Technologies, Inc.'s Telecommunications and Information Task Force.

David J. Cowen is head of the Geography Department, codirector of the NASA Visiting Investigator program, and Carolina Distinguished Professor of Geography at the University of South Carolina. His research and instruction have focused on geographic information systems. He chairs the Mapping Science Committee of the National Research Council. Dr. Cowen earned B.A. and M.A. degrees from the State University of New York at Buffalo and a Ph.D. degree in geography from Ohio State University.

Donna P. Koepp is currently head of the Government Documents and Microforms Library and head of reference and instruction for the Social Sciences program at Harvard University. Her specialty has been government-produced maps and the history of mapping by federal agencies. Ms. Koepp is active in the Cartographic Users Advisory Council, representing the American Library Association Government Documents Round Table and acting as liaison to the U.S. Government Printing Office and the Federal Geographic Data Committee. Following a B.A. degree from the University of Colorado, Boulder, she worked at the Denver Public Library with government documents and maps as a subject specialist and reference librarian and received an M.A. in librarianship from the University of Denver. Ms. Koepp worked with maps and government documents at the University of Kansas before moving to Harvard University.

Xavier R. Lopez is director of Oracle's Location Services group. Dr. Lopez leads Oracle's efforts to incorporate spatial technologies across Oracle's database, application server eBusiness applications. He has 12 years of experience in GIS and spatial databases. He holds advanced engineering and planning degrees from the University of Maine, Massachusetts Institute of Technology, and the University of California, Davis. Dr. Lopez has been active in numerous academic and government research initiatives on geographic information. He is the author of a book on government spatial information policy and has written over 50 scientific and industrial publications in areas related to spatial information technology. Since December 2001 Dr. Lopez has served on the Committee on Multimodal Transportation Requirements for Spatial Information Infrastructure at the National Research Council.

Richard D. Miller is the Kansas chief information technology architect and director of the Kansas Information Technology Office. These duties encompass coordination of enterprise information technology activities, including strategic planning, policy development and implementation, and IT architecture development and maintenance. He is also director of the Kansas Geographic Information Systems Initiative, serving in this role since 1995. Previous experience in spatial and information technologies include that of supervisory geographer for the U.S. Bureau of the Census and as GIS coordinator and chief of systems management for the Kansas Department of Health and Environment. Mr. Miller represents the GIS community on the Kansas Information Technology Advisory Board, serves as secretary to the Kansas Information Technology Executive Council, and is president of the National States Geographic Information Council. He is also former consortium and symposium chairman for the Mid-America Geographic Information Consortium. He holds a B.S. in education and an M.A. in geography, both from the University of Kansas.

Gale W. TeSelle recently retired from the U.S. Department of Agriculture, Natural Resources Conservation Service, where he was the national GIS program manager, and the acting chief information officer of the Natural Resources Conservation Service. Mr. TeSelle received his bachelor's degree in geography from the Nebraska Wesleyan University and his Master's degree in geography from the University of Nebraska. Mr. TeSelle was a founding member of the National Aerial Photography program, National Digital Orthophotography program, and the Federal Interagency Coordinating Committee on Digital Cartography. As chair of the latter's Standards Working Group for seven years, he authored the National Geo-Data System plan, which served as the concept and vision for the National Spatial Data Infrastructure. Mr. TeSelle helped form the Federal Geographic Data Committee and served as an active member of the FGDC Coordination Committee.

Waldo R. Tobler received his degrees in geography from the University of Washington in Seattle, and is currently professor emeritus of geography at the University of California at Santa Barbara. Dr. Tobler was one of the principal investigators and a senior scientist in the National Science Foundation-sponsored National Center for Geographic Information and

Analysis. He has served on numerous entities of the National Research Council, the most recent being the Board on Earth Sciences and Resources. He has been on the editorial board of several journals, including *The American Cartographer*, *Journal of Regional Science*, *Geographical Analysis*, and the *International Journal of Geographical Information Systems*. He is a member of the National Academy of Sciences and until his retirement was a member of the Royal Geographical Society of Great Britain. He served as the U.S. delegate to the International Geographical Union commission on geographical data processing and sensing. Current concerns relate to ideas in computational geography, including the analysis of vector fields and the development of global trade models.

Nancy von Meyer, vice-president of Fairview Industries, provides consulting, education, and GIS implementation services to government agencies and the private sector. She received her Ph.D. in civil and environmental engineering from the University of Wisconsin-Madison in 1989. Dr. von Meyer works with many counties and local governments on parcel, land records, and system design for automation and modernization projects. She is also active in federal initiatives related to the FGDC Cadastral Data Content Standard, the National Integrated Lands System, eastern states cadastral initiatives, and other land records projects. Dr. von Meyer served on the Mapping Science Committee of the National Research Council from 1994 to 1997.

National Research Council Staff

Paul M. Cutler, *study director,* is a program officer for the Board on Earth Sciences and Resources of the National Research Council. He directs the Mapping Science Committee and ad hoc studies on earth science and mapping science issues. Before joining the NRC Dr. Cutler was an assistant scientist and lecturer in the Department of Geology and Geophysics at the University of Wisconsin-Madison. His research is in the area of surficial processes, specifically glaciology, hydrology, and quaternary science. In addition to numerical modeling and GIS research he has conducted field studies in Alaska, Antarctica, arctic Sweden, the Swiss Alps, Pakistan's Karakoram mountains, the midwestern United States, and Canadian Rockies. He is a member of the Geological Society of America, American Geophysical Union, Geological Society of Washington, and a fellow of the Royal Geographical Society. He received a bachelor's degree from Manchester University, England, a master's degree from the University of Toronto, and a Ph.D. from the University of Minnesota.

Eileen M. Mctague is a research assistant for the Board on Earth Sciences and Resources of the National Research Council. She holds an M.S. degree in environmental science from American University and a B.S. degree in biology from Pennsylvania State University. Ms. McTague has interned at the National Academy of Engineering, the Renewable Natural Resources Foundation, and Discovery Creek Children's Museum.

Radhika S. Chari is a senior project assistant for the Board on Earth Sciences and Resources of the National Research Council.

Appendix B

Oral and Written Contributions

Hugh Archer,[*] Kentucky Department of Natural Resources
Ernest Baldwin,[*] U.S. Government Printing Office
Yves Belzile,[*] Natural Resources Canada
Hugh Bender,[*] Texas Natural Resources Information System
Tom Berg, State Geologist, Ohio
Ronald Birk, National Aeronautics and Space Administration
Scott Cameron,[*] U.S. Department of the Interior
Donald Cooke,[*] Geographic Data Technology
William Craig,[*] University Consortium for Geographic Information Science
Kari Craun, U.S. Geological Survey
Patti Day,[*] American Geographical Society Library, University of Wisconsin-Milwaukee
Drew Decker, Texas Strategic Mapping Program
Brook DeLorme, DeLorme Maps
Michael Domaratz, U.S. Geological Survey
Shoreh Elhami,[*] Delaware County Auditor's Office, Ohio
Jeanne Foust,[*] Environmental Systems Research Institute

---

[*]Indicates a white paper submission to the committee. White papers may be found at: < http://www7.nationalacademies.org/besr/National_Map_Participants. html >.

Patrick Fowler, Environmental Systems Research Institute
Dennis Goreham,* State of Utah, Automatic Geographic Reference Center
Charles E. Harne,* Bureau of Land Management
John Hebert, Library of Congress
Richard Hogan, U.S. Geological Survey
Bruce Joffe,* Urban and Regional Information Systems Association
Richard Kleckner, U.S. Geological Survey
Scott Loomer, National Imagery and Mapping Agency
Michael Mahaffie,* Delaware Office of State Planning Coordination
Robert Marx,* U.S. Bureau of the Census
Anne Hale Miglarese,* National Oceanic and Atmospheric Administration
Scott Oppmann,* Oakland County, Michigan
John Palatiello,* Management Association for American Private
    Photogrammetric Surveyors
James Plasker,* American Society for Photogrammetry and Remote
    Sensing
Stanley Ponce, U.S. Geological Survey
Mark Reichardt, Open GIS Consortium
J. Milo Robinson,* Federal Geographic Data Committee
Paul Rooney, Federal Emergency Management Agency
Barbara Ryan, U.S. Geological Survey
Rebecca Somers, Somers-St. Clair
Curt Sumner,* American Congress on Surveying and Mapping, Inc.
Eugene Trobia,* National States Geographic Information Council
Ian von Essen,* Washington State Geographic Information Council
John Voycik,* Greenhorne & O'Mara, Inc.
James Wescoat, National Imagery and Mapping Agency

---

*Indicates a white paper submission to the committee. White papers may be found at: < http://www7.nationalacademies.org/besr/National_Map_Participants. html >.

Appendix C

Previous Recommendations and Observations on National Mapping Activities

TABLE C.1 Previous recommendations and observations sampled from reports and meetings providing advice on national mapping activities.

Date	Report/Meeting—Sponsor/Author
1980	**Need for a Multipurpose Cadastre—NRC (1980)**

"Numerous conferences have been held and reports prepared that discuss the problems with our present land-information systems, such as duplication, lack of accessibility, single-purpose data systems, lack of standards, and institutional arrangements that limit coordination among land—related functions."

Selected Recommendations:

1.   That local governments maintain land data compatible with a multipurpose cadastre and transmit these data to high levels of government when needed.

2.   That federal legislation be prepared to authorize and fund a program to support the creation of a multipurpose cadastre in all parts of the nation.

TABLE C.1 Continued

Date	Report/Meeting—Sponsor/Author

3. That the Office of Land Information Systems established by each state ... be responsible for:

- Promoting effective, efficient, and compatible land-information systems among governmental levels, in cooperation with the federal government to ensure compatibility on a national basis;
- Setting standards for state, regional, and local government surveying, mapping, and land-data collection efforts, making use of federal technical studies;
- Providing guidance to those local offices with major responsibilities for land information (namely, recorders, assessors, surveyors, engineers, and planners);
- Serving as the focal point and clearinghouse for state and federal agencies collecting or mapping land information, taking responsibility for the quality of the information that is forwarded; and
- Enlisting the resource of other state agencies having important contributions to make to the developments of the cadastral system, especially those responsible for land assembly, construction, and management of public lands and efficiency of state administrative services.

1981   **Federal Surveying and Mapping: An Organizational Review—NRC (1981)**

1. The Panel believes that state and local governments should assume greater responsibility for those surveying and mapping functions that relate to regional and local programs. However, the federal, state, and local roles must be completely integrated.

2. We recommend that the mapping, charting, geodesy, surveying, and cadastral agencies of the federal government continue to sponsor cooperative programs, with state and local governments providing sufficient guidance to ensure conformance to national specifications and standards and thus to the development of a fully integrated national information system.

3. We recommend that the geodetic and cartographic data bases be adequately supported, be readily accessible to all users, and, even though serving different interests and needs, be made integral parts of a national mapping, charting, geodesy, surveying, and multipurpose cadastre information system.

Date	Report/Meeting—Sponsor/Author

1982 Modernization of the Public Land Survey System—NRC (1982)

1. This committee endorses the creation of a Federal Surveying and Mapping Administration. In the interim, we recommend an interagency working group be formed with participation of all relevant federal agencies and interested groups at the state, local, and private sector levels to integrate the geodetic, cadastral, and mapping activities necessary for the modernization of the Public Land Survey System.

2. It is recommended that the USGS National Mapping Division accelerate the development of digital files of the 7.5 minute mapping series and provide documentation of the current status, cost, and methods and procedures being employed in the development of the National Digital Cartographic Data Base.

1982 Duplicative Federal Computer Mapping Programs: A Growing Problem—GAO (1982)

1. GAO recommends that the Director of OMB issue a circular or other directive requiring the interagency coordination of computer mapping and preventing duplicative programs.

2. Interior to accelerate the production of computerized maps most in demand by other federal agencies.

3. OMB agreed that it should take action to improve the coordination of federal computer mapping.

1990 Circular A-16 Revised—OMB (1990)

1. The major objective is the eventual development of a national digital spatial information resource with the involvement of Federal, State, and local governments, and the private sector. This resource would include base topographic mapping, cadastral, geologic, geodetic, soils, wetlands, vegetation, cultural, demographic, and ground transportation data.

2. Coordinate Federal survey, mapping, and related spatial data activities and establish the FGDC as the coordination mechanism.

3. Promote the development, maintenance, and management of distributed spatial databases that are national in scope.

TABLE C.1   Continued

Date	Report/Meeting—Sponsor/Author
1990	**The Future of the National Mapping Program—NRC (1990)**

1.   Expand the role in developing the National Digital Cartographic Data Base (NDCDB) so that its functions include management and coordination, standards setting and enforcement, data production, cataloging, and data dissemination and related services.

2.   Increase its activities to provide a large number of classes of spatial data to better meet national needs both within the earth science/ natural resources sector and other sectors that are dependent on spatial data.

3.   Speed the creation of the NDCDB by increasing emphasis on work-sharing and cost-sharing programs, developing prototyping, testing and implementing a digital data donor program throughout the public and private sectors, and allocating adequate National Mapping Division resources to information management and user/donor coordination, and, if necessary increasing these relative to traditional data production programs.

4.   Continue and, if possible, expand its efforts in establishing and promulgating digital spatial data quality standards, including standards for larger-scale data sets and maps.

5.   Establish plans for and begin prototyping a national spatial database, which would be an enhancement of the NDCDB and would be feature-oriented and on-line by the year 2010 or sooner.

6.   Expand its current research activities in digital cartography, geographic information systems and remote sensing, and image processing.

7.   Transform USGS's National Mapping Division from a mapping service organization to the federal agency responsible for structuring and coordinating the geographic or spatial component of the national infrastructure.

| **1993** | **Toward a Coordinated Spatial Data Infrastructure—NRC (1993)** |

Effective national policies, strategies, and organizational structures need to be established at the federal level for the integration of national spatial data collection, use, and distribution.

Date	Report/Meeting—Sponsor/Author

1. Procedures should be established to foster ready access to information describing spatial data available within government and the private sector through existing networks, providing on line access by the public in the form of directories and catalogs.

2. A spatial data-sharing program should be established to enrich national spatial data coverage, minimizing redundant data collection at all levels, and create new opportunities for the use of spatial data throughout the nation. Specific funding and budgetary crosscutting responsibilities of federal agencies should be identified by OMB, and the FGDC should coordinate the crosscutting aspects of the program.

1993   **Creating a Government that Works Better and Costs Less—Report of the U.S. National Performance Review (1993)—Vice President Gore**

1. In partnership with state and local governments and private companies, we will create a National Spatial Data Infrastructure.

2. The administration will develop an NSDI to integrate all of the data sources into a single digital resource accessible to anyone with a personal computer.

3. The FGDC, which operates under the auspices of the OMB, plans to raise enough non-federal funding to pay for a least 50 percent of the project's cost.

1994   **Promoting the National Spatial Data Infrastructure Through Partnerships—NRC (1994)**

1. The size and diversity of the federal establishment suggest that viable partnerships will require focal points within the federal government for coordinating data production and partnerships activities.

2. Clear guidelines for cost sharing in partnerships need to be developed.

3. It is imperative that state and other organizations be involved in the standards development process and that only standards essential to NSDI objectives be required of partnership agreements.

4. Incentives are needed to encourage partnerships that are designed to maximize use and benefits to the broader user community.

TABLE C.1   Continued

Date	Report/Meeting—Sponsor/Author

5. The FGDC should investigate the extent to which federal procurement rules (and future revisions resulting from the National Performance Review) are an impediment to the formation of spatial data partnerships and identify steps that can be taken to ease them.

1994   Executive Order 12906—President Clinton

1. The Executive Order is intended to strengthen and enhance the general policies described in OMB Circular A-16. Each agency shall meet its respective responsibilities.

2. In consultation with State, local, and tribal governments and within nine months of the date of this order, the FGDC shall submit a plan and schedule to OMB for completing the initial implementation of a national digital geospatial framework by January 2000. At a minimum, the plan shall address how the initial transportation, hydrology, and boundary elements of the framework might be completed by January 1998 in order to support the decennial census of 2000.

3. The Secretary, under the auspices of the FGDC, and within 9 months of the date of this order, shall develop, to the extent permitted by law, strategies for maximizing cooperative participatory efforts with State, local, and tribal governments, the private sector, and other nonfederal organizations to share costs and improve efficiencies of acquiring geospatial data consistent with this order.

1995   A Data Foundation for the National Spatial Data Infrastructure — NRC (1995)

1. The Mapping Science Committee (MSC) recommends that geodetic control, orthorectified imagery, and terrain (elevation) data be considered the critical foundation of the NSDI.

2. The FGDC should be responsible for coordinating the development and certification of a foundation and for its maintenance and availability. Programs to acquire the data that comprise the foundation should be accelerated to ensure that the foundation is adequate to meet the needs of the NSDI, particularly for the integration of other data. Data partnerships among federal agencies, state and local governments, the private sector, and others should be a key component of these programs.

Date	Report/Meeting—Sponsor/Author

3. Specific spatial data themes should be designated as Framework data.

4. The FGDC should:

   ▪ Coordinate identification of the various components of existing framework data through its clearinghouse;
   ▪ Encourage efforts to integrate those data with the foundation; and
   ▪ Identify gaps in data coverage and encourage the establishment of programs that include partnerships to populate these framework data themes.

5. To accomplish the needed compilation, maintenance, quality control, and access of the foundation and framework data, additional research and development efforts are required to technically support these activities.

6. The MSC also suggested that, at a minimum, Framework data must:

   ▪ Be compiled, archived, and maintained in digital form;
   ▪ Include metadata descriptions;
   ▪ Be mathematically and semantically integrated with the NSDI foundation;
   ▪ Be available in an accepted, openly publicized, standard data exchange format; and
   ▪ Be accessible to the public.

7. The TIGER files could be integrated with the foundation by the following actions:

   ▪ Improving coordinate accuracy using orthorectified imagery that is tied to the geodetic control network;
   ▪ Completing and improving street and address coverage in partnerships with the U.S. Postal Service, 911 emergency agencies, state and local government, and the private sector; and
   ▪ Establishing an ongoing update facility employing local government partnerships for timely information (transactional updates) about new streets.

TABLE C.1 Continued

Date	Report/Meeting—Sponsor/Author
1996	**Congress on Applications of GIS to the Sustainability of Renewable Natural Resources—Renewable Natural Resources Foundation (1996)**

1. Despite the wealth of data that already exists, we still do not have a good understanding of the condition of our natural resources.

2. Federal agencies should take the lead in developing baseline data sets.

3. We need better metadata.

4. Simplify the use of GIS through Web-based interfaces.

5. Federal, state, and local governments, nongovernmental organizations, and the private sector must work to improve cooperation and coordination.

6. We need more commitment to use existing mechanisms of coordination such as the Federal Geographic Data Committee.

7. We need better public/private partnerships.

1997   The Future of Spatial Data and Society—NRC (1997)

While the report does not include specific recommendations, it does list a number of trends that will impact the NSDI, framework, and *The National Map*, including:

- Analysis, visualization, and cognitive technologies: Development of search and integration tools, and virtual reality;
- Pervasiveness of technology: Improved access to data and technology;
- Data Integration: Spatial data becomes transparent to the user;
- Timely data and use: Needs for currency will change data management;
- Quality assurance/quality control: Greater role of metadata;
- Spatial literacy: Increased geo-understanding;
- Partnerships: Data utilities may emerge;
- Control of data: Conflicting public policies will continue;
- Data collection agents: More local collection of spatial data;

Date	Report/Meeting—Sponsor/Author

- Data security and protection: Possible restriction on public access;
- Decision-making process: Spatial analysis continues to grow in importance;
- Citizen involvement: Spatial capabilities will expand involvement; and
- Privatization: Suggestive trend of increasing privatization.

1997   NSDI Framework, Introduction and Guide—FGDC (1997a)

1.   Framework is built through cooperative efforts. Partnerships and cooperation for geographic data-sharing activities among local, state, and federal governments and the private sector are essential for the development of framework.

2.   The goal is to have nationwide framework data coverage composed of the pieces produced for smaller area extents.

3.   Framework data are seamless across collection areas.

4.   Framework data are consistent among themes.

5.   Framework data supports transactional updates.

6.   Charges for access to framework are limited to the costs of providing access or dissemination.

7.   Framework contributions by contributors must be certified.

8.   Develop a good business plan for building framework data.

9.   Clearly identify components of framework.

10. Provide incentives for developing framework.

11. Near-term needs necessitate a phased implementation that will put useful pieces in place as soon as possible.

12. Specific goals in phase two (1995-98) include implementing institutional arrangements for framework and begin to reorient federal agency and other program activities to support the framework.

TABLE C.1   Continued

Date	Report/Meeting—Sponsor/Author

13. The goal in phase 3, beginning in 1998, is to spread framework participation and bring the framework to maturity.

1997   A Strategy for the NSDI—FGDC (1997b)

1.   Increase the awareness and understanding of the vision, concepts, and benefits of the NSDI through outreach and education.

- Demonstrate the benefits of participation in the NSDI to existing and prospective participants;
- Promote principles and practices of the NSDI through formal and informal education and training; and
- Identify and promote the attitudes and actions that help to develop the NSDI.

2.   Develop common solutions for discovery, access, and use of geospatial data in response to the needs of diverse communities.

- Continue to develop a seamless National Geospatial Data Clearinghouse;
- Support the evolution of common means to describe geospatial data sets;
- Support the development of tools that allow for easy exchange of applications, information, and results; and
- Research, develop, and implement architectures and technologies that enable data sharing.

3.   Use community-based approaches to develop and maintain common collections of geospatial data for sound decision-making.

- Continue to develop the National Geospatial Data Framework;
- Provide additional geospatial data that citizens, governments, and industry need;
- Promote common classification systems, content standards, data models, and other common models to facilitate data development, sharing, and use; and
- Provide mechanisms and incentives to incorporate multi-resolution data from many organizations into the NSDI.

4.   Build relationships among organizations to support the continuing development of the NSDI.

Date	Report/Meeting—Sponsor/Author

- Develop a process that allows stakeholder groups to define logical and complementary roles in support of the NSDI;
- Build a network of organizations linked through commitment to common interest within the context of the NSDI;
- Remove regulatory and administrative barriers to agreement formation;
- Find new resources for data production, integration, and maintenance;
- Identify and support the personal, institutional, and economic behaviors, and technologies, policies, and legal frameworks that promote the development of the NSDI; and
- Participate with the international geospatial data information community in the development of a global data infrastructure.

1997   **Geospatial Information Infrastructure Master Plan—NIMA (1997)**

1.   Change the geospatial support strategy to provide the geospatial information framework for an integrated and interoperable view of the mission space.

2.   Change the NIMA production strategy to provide a near-global foundation of geospatial information that can be intensified to meet the requirements for mission specific data sets, including NIMA standard products.

3.   Update the NIMA mission to incorporate the development and maintenance of a shared geospatial framework of information and services.

4.   Implement a new requirements process that ties together user missions, resources, and systems with the essential elements of geospatial information needed for operational success.

5.   Develop a web-based architecture to facilitate access to the share framework and to speed the dissemination of framework information and services.

6.   Equip and train users to exploit new resources.

7.   Establish the infrastructure and business practices needed to sustain the changes.

TABLE C.1 Continued

Date	Report/Meeting—Sponsor/Author
1998	**Geographic Information for the 21st Century—Building a Strategy for the Nation—NAPA (1998)**

1. Interagency, intergovernmental, and private sector GI users and producer groups, whose cooperation is essential to implementing NSDI, should continue to be convened to encourage and accelerate the development, sharing, and maintenance of NSDI framework data files. These groups should be used to negotiate additional data sharing and joint funding agreements.

2. The federal government policy of promoting open access, especially for all data used in public policy decision-making, should be maintained and the state and localities should be urged to adopt similar policies.

3. Establish through legislation a national goal to create and maintain a robust NSDI.

4. Forward to Congress legislation to transfer the National Geodetic Survey to USGS and authorize the establishment of a Geographic Data Service.

5. Develop a reorganization plan to implement the GDS and realign the federal field structure for [the development of] base geographic information.

6. Develop coordinated goals, strategies, performance measures, and budgets for federal agency GI programs and activities. Explicitly establish selected strategic goals and performance measures, as required by the Results Act, to help move the NSDI toward further and faster realization.

7. The FGDC should act as the focal point for coordinating the high priority GI technology needs of civil government at all levels and for mobilizing interagency, state, and local support for selective high-payoff technology developments with utility in multiple civil applications.

8. NIMA should become much more actively engaged in the FGDC because of the increasing need to coordinate GI activities, including

Date	Report/Meeting—Sponsor/Author

technological research, standards, security policy, procurement practices, and international activities.

1999   **FGDC Coordination Retreat at Shepherdstown, W. Va.—FGDC (1999a)**

1. Create and adopt a common vision for "Framework."

2. Develop content standards for framework data.

3. Improve the integration of framework data themes.

4. Foster collaborative data collection and integration of framework data.

5. Improve agency compliance and involvement in framework data development.

6. Encourage all lead agencies to develop a framework action plan.

7. Sell NSDI benefits to Congress and other agencies.

8. Sponsor a 2001 cross-cut initiative for framework data.

9. Establish performance measures for NSDI goals.

10. Develop public/private partnerships through the use of a web portal that includes FGDC Clearinghouse and commercial webpages.

1999   **FGDC Steering Committee/Stakeholders Meeting—FGDC (1999b)**

1. Framework must be built.

2. Must have strong local, state, and federal involvement.

3. Enact policies that define the roles in managing the framework data.

4. Provide leadership.

5. Sustain funding support for framework activities.

TABLE C.1 Continued

Date	Report/Meeting—Sponsor/Author

1999   **National GeoData Forum — FGDC (1999c)**

Questions raised at the meeting and issues discussed relevant to the topic of this report:

1. How can the current investment in framework data be leveraged to maximize benefits and minimize cost?

2. What are the barriers to achieving the vision to develop, maintain, and integrate framework data within a geographic area?

3. How can the private sector contribute to and benefit from this effort?

2000   **Improving Federal Agency Geospatial Data Coordination— FGDC (2000)**

1. Get serious about coordination among federal agencies.

2. Use FGDC funds to provide partial support for a GI coordinator in key federal agencies.

3. Foster partnerships among federal agencies by focusing on specific applications requiring interagency sharing and coordination of geospatial data and budget initiatives.

4. Promote other federal agency support and testimony at FGDC member agency budget hearings.

5. Re-evaluate and redefine framework and the federal role in its development.

6. Develop a model of multi-agency data development and user participation that includes local, state, and national players.

7. Develop a coordinated crosscut budget initiative to achieve the vision of integrated geospatial data assets.

8. Bring agencies together to facilitate development of national data assets and complementary standards.

Date	Report/Meeting—Sponsor/Author

9. Promote the initiation of interagency collaborative projects.

10. Develop an implementation plan with timelines and responsible parties once the recommended actions are adopted. Review the action plan with OMB and add appropriate progress measures for performance management

2000   **Geospatial Information Roundtable—OMB (2000)**

1. We need to develop data assets as a nation and accelerate framework data and standards.

2. Much work still is needed on data standards.

3. Our challenge is to leverage investments that we are already making and budget for the full life cycle.

4. We need to look at financial strategies that include more than appropriations.

5. The primary barriers to success are organizational and financial.

6. We need to empower state, local, tribal, local federal and others to build consistent framework data layers.

7. There is not an easy way to track the status of framework. We need to build and maintain a comprehensive NSDI framework inventory.

8. Although a policy framework exists, agencies have had mixed success in overcoming the technical, institutional, and financial barriers to developing an NSDI that is fully populated with current, accurate, and readily accessible spatial data.

2001   **National Spatial Data Infrastructure Partnership Programs: Rethinking the Focus—NRC (2001)**

This report did not include the typical set of recommendations. It reviewed the FGDC partnership programs and provided the following suggestions:

1. The partnership programs should have more rigor so they can be a true assessment of whether the funds have made a difference.

TABLE C.1 Continued

Date	Report/Meeting—Sponsor/Author

2. The partnership programs need to take a long-term view. One year funding is inadequate to assess the real success.

3. The partnerships should be established in a hypothesis testing mode so that there are measurable outcomes.

4. The MSC also suggested the need for an extended framework.

5. It is becoming increasingly obvious that an effective and widely used NSDI will be developed with substantial if not primary input from organizations outside of federal government.

6. The requirement for large-scale source materials is critical for the development of federal-local partnerships.

7. In other words, if a data layer is part of the NSDI and also a component of both a State Spatial Data Infrastructure (SSDI) and a Local Spatial Data Infrastructure (LSDI), the data for these layers need to be collected at the lowest level and generalized to the other levels. This ensures logical consistency among the parts of the extended NSDI framework.

8. There are at least nine major steps necessary to realize this extended Framework:

- Definition of the contents of the city, county, or local extended Framework;
- Definition of the contents of the state or tribal nation extended Framework;
- Definition of the extended Framework hardware architecture;
- Definition of coordination mechanisms;
- Assignments for layer responsibilities;
- Definition of quality standards (collection and maintenance) and procedures for the development of the extended Framework at all levels;
- Data generation in agreement with the corresponding Framework;
- Data maintenance program; and
- Budget allocation.

Appendix D

Selected Points Raised in Written Comments of Workshop Participants[1]

Hugh Archer
Kentucky Department for Natural Resources

- The vision for *The National Map* is, although not particularly new, quite inspirational. It embodies a direction that NMD and the many potential partners must incorporate and move toward. The vision, however, is not matched to the proposed timetable, realistic budget expectations, a rational implementation plan, and is particularly lacking in any plan to make the necessary partnerships a reality.

- Most GIS users in the country can create useful base maps right now without the national program, although the *National Map* vision would certainly expand access to a better base map than many nonprofessionals might have access to today. Today's broader range of users and possible sources for geospatial data merely complicate the institutional arrangements necessary to take advantage of the growing but disjunct data resources at some national seamless level. A growing number of different resolutions and scales are part of the data resources that would constitute and update *The National Map* in concept. Necessary standards

---

[1]Full documents are available at <http://www7.nationalacademies.org/besr/ National_Map_Participants.html>.

of all types and classifications enumerated in many papers still are only concepts, not a reality of practice.

▪ When terrorism strikes again the best preparation (outside of counterintelligence efforts) will be to create a system of high-quality data that is created locally and data mined up through the states and into the federal network. Any value-added items and funding should be sent back down the network. To save lives this data must be sitting on both federal and local authorities' desks, and they must have it in use.

▪ The effort needs to be recognized for how important to all forms of cooperation, communication, and cost-effective services it will be. Perhaps we will see the states, the utilities, the home security institutions, and the private sector associations standing together with USGS explaining how it needs to work to Congress and OMB in the near future.

Ernest Baldwin
U.S. Government Printing Office
_____

▪ Keep the interface simple and readily available to the general public user. If possible, use a standard Web browser interface. To minimize undue complexity, maintenance, and expense, proprietary client software and other products with copyright-like barriers should be avoided. Design *The National Map* for full functionality over typical connection bandwidths and speeds. *National Map* designs should target "middle-of-the-road" personal computer, monitor, and printer hardware. Avoid the necessity for bleeding-edge technologies. A compromise may be to design tiered functionality, offering a basic set of functions to lower-end users.

▪ *The National Map* should be accessible to visually impaired users; i.e. (USGS should design in Section 508 compliance).

▪ The FDLP and the Cataloging and Indexing program work in concert to provide timely, permanent, no-fee public access to U.S. government publications. This is achieved through the operation of a network of libraries that contain collections of U.S. government publications and provide services to assist the public in using this material; creation and maintenance of tools to identify, describe, locate, and obtain publications; and maintenance of permanent collections of U.S. government publications.

▪ U.S. government maps are a significant component of the FDLP, and FDLP distribution of USGS printed maps is accomplished by USGS for GPO, operating under the terms of an interagency agreement. This partnership began in the mid-1980s. Since 1996 GPO has been acting under congressional direction to emphasize the use of online information in the FDLP, and today over 60 percent of the new titles in the program are disseminated on the Internet.

- There is considerable demand for cartographic products in the FDLP. Many depository libraries have deep historical collections of USGS printed maps. The National Atlas maps were selected by over half of the libraries, so it is reasonable to assume that *The National Map* will receive considerable use from this sector.

Yves Belzile
Natural Resources Canada

- The Canadian GeoBase initiative suffers from an "an all things to all people" flavor that generates both unreasonable expectations with the users and skepticism from the suppliers.
- The access to *The National Map* data and functionalities need to be easy if we expect the users to accept the operating environment and work from within. The advantages of joining the crowd are undeniable as long as you are capable and can afford doing so.

Hugh Bender
Texas Natural Resources Information System (TNRIS)

- One issue to bring forward for this summit is that *The National Map* in concept has been occurring for a number of years within many states. The question is whether the federal government capitalizes on these efforts and assumes a cooperative leadership role in the development of data and coordination of data assimilation. An important change associated with *The National Map* is that the USGS as the national lead reorganizes itself into an organization that supports the efforts of other agencies, especially federal, in this complex web of coordination and partnerships.
- In the 21st Century Volunteer Support for the National Map of Texas comes [in the] form of two organizations associated with TNRIS. The first is the Texas Geographic Information Council (TGIC), a group of 40+ local, regional, and state and federal agencies coordinating the development and use of geospatial data and technology in Texas. The second group is the recently created nonprofit corporation, Texas

Geographic Society, or TxGS. This public-private society is the catalyst for a number of essential elements of *The National Map*.

- Technical issues will not decide the success of *The National Map*, as these are rapidly being resolved throughout this country. Success on a federal level will be achieved through attitudes of cooperation, shared responsibility, and shared or match funding.

- If the federal government is serious about homeland security then it requires *The National Map*. Ironically, the best justification for *The National Map* comes from the sudden interest in homeland security and the obvious benefits for disaster response and prevention. The intelligence community, with the best data at its disposal (local) will prevent many events as it has in the past. *The National Map* is a chance to increase its success rate by providing higher-quality and consistent data themes at its disposal.

- Technical challenges include cooperation and partnerships but understanding your customer and consistently making the data easier to use will make the map successful. The experience and approach taken by TNRIS technical staff is based on everyday experience with thousands of regular and new users that cover the spectrum of experience and knowledge of GIS and computers. Recognizing the exponentially growing need by an ever-varied segment of the population means that the time for a *National Map* has come. Society today is accustomed to receiving accurate information with the click of a mouse 24-7-365, and they assume it is current.

Scott Cameron
U.S. Department of the Interior

- Biggest challenges will be (1) coordination among hundreds (thousands) of participants, and (2) developing incentives for state and local governments to share/contribute their data.

- The Geospatial One-Stop project's "data acquisition bazaar" may be a useful mechanism by creating a national market for data acquisition, encouraging the economically most efficient allocation of federal, state, and local financial resources.

Donald Cooke
Geographic Data Technology (GDT)

- It is worth noting that the requirement for getting new addresses into master street address guides, for dispatching emergency services, is within 48 hours of the phone being hooked up. While the technical chores (identifying the fire, police and ambulance zones for a new structure) are simpler than adding a new street to a vector database, this requirement has been routinely achieved for years.
- As Jack Dangermond said a couple of years back, "We're moving from months and years (of currentness) to minutes and seconds." You can't keep a spatial database updated by replacement and expect the kind of currentness today's applications require. You have to use transactions. The GIS software folks will have to implement this part of the vision. This is a big deal and won't be easy; we've got to do it.
- More GIS users need current data than most people believe: Fedex and UPS deliver to contractors on building sites. Waste Management delivers and picks up the dumpsters. Sears Logistics delivers and installs the dishwashers. Workers sustain injuries and require ambulance services. All of this happens before anyone moves in on a new street. This isn't some future vision; this has been common GIS usage for years.
- Feature-level metadata is more of a "footnote" metaphor: data about data down at the feature, or attribute, level. The node coordinates at one end of a street vector may have been GPS'ed to 1-meter accuracy, while the coordinates at the other end may have come from a 1:100,000 DLG. This is crucially important information, which must be stored for each node. We've been doing this for years at GDT, and I don't see how we could function without our "confidence codes."

William Craig
University Consortium for Geographic Information Science (UCGIS)

- It is clear that significant coordination and cooperation will be required to weave this data into a whole cloth. One of the major challenges facing this effort is getting the enthusiastic cooperation of state and local government to share their data, especially data they are now licensing and selling to provide income to cover the costs of data collection and maintenance. State and regional coordinating bodies can assist with this effort, but challenges remain.

▪ A few [research issues] are critical to the success of this effort. *Spatial Data Acquisition and Integration* ... are at the heart of the concept, yet continue to need research and development to yield the desired product. *Geographic Information Engineering* ... includes interoperability and mobile computing, issues that are cutting edge for industry, government, and this initiative. *GIS and Society* ... includes the institutional aspects that are the basis for sharing data. UCGIS scientists have already made significant progress on many of these issues. We are excited by the prospect of having our work contribute to the realization of *The National Map.*

Patti Day
American Geographical Society Library, University of Wisconsin-Milwaukee

▪ USGS should work with the FGDC and Geospatial One-Stop to facilitate the alignment of roles, responsibilities, and resources at the state and local levels. USGS should work with Geospatial One-Stop and I-Teams on multisector coordination, development, and implementation of standards to create the consistency needed for interoperability.

Shoreh Elhami
Delaware County Auditor's Office, Ohio

▪ There isn't enough mention of local government's role in general. Even the previous meetings and workshops have not included as much local government representation as it should have. . . . Since there will be a lot of requests made to local governments for datasets, as customized data and applications become available, would the locals have to pay for those add-on services? It would make sense to make these services available to them free of change as an incentive so they would continue their cooperation with USGS. . . . Why not use the state or regional consortiums instead of adding another level of bureaucracy. . . . Generally, the question is, what is USGS's strategy when it comes to changes in government (as a result of elections) and its impact on cooperative agreements?

Jeanne Foust
Environmental Systems Research Institute (ESRI)

- *The National Map* will be the foundation on which a number of additional data layers are built that are vital to the private sector. For example, the flood plain maps produced by FEMA are directly based on the digital elevation data created by topographic map series from USGS. Insurance companies use the flood plain plus topographic data for insurance underwriting. Banks must include a flood plain data marker for every building mortgage they fund. Census geographic and associated demographic data for retail site selection and market area planning are also tied to a national base map.

Dennis Goreham
State of Utah, Automatic Geographic Reference Center

- *The National Map* must have measurable outcomes/performance measures that include consequences for nonperformance.
- To encourage broad-based participation, *The National Map* must be easy to understand, easy to participate in, and have obvious benefits for all stakeholders.

Charles E. Harne
Bureau of Land Management

- A necessary part of these [*National Map*] partnerships will be acknowledging local requirements and specifications, at the expense of a national standard. Because of different requirements and budgets, datasets resulting from these partnerships will be an eclectic and overlapping assortment of elevation data and of imagery data. . . . It will be tempting to normalize raster coverages so that they offer a more manageable serving environment. With this approach overlapping coverages would not be supported. . . . It may be preferable to store all coverages at full resolution and retain their unique attributes. Likewise, overlapping coverages should be available to provide the user the ability to specify their data preferences. . . . Exactly how this capability can be developed requires additional investigation. Nothing described above is beyond the capability of existing technology. Various strategies, such as cascading servers could be employed to distribute "on-the-fly" processing. In any case *The National Map* program will have to evaluate user requirements for serving raster data from different sources and multiple coverages and arrive at a serving strategy that emphasizes the user and not the system.

Mike Mahaffie
Delaware Office of State Planning Coordination

- An uncoordinated approach to the development and use of spatial data wastes taxpayer money and reduces the value of information generated by the use of that data. It is wasteful and duplicative for different agencies and levels of government to invest time and money in the creation and maintenance of the same datasets.

- It is important that *The National Map* effort tie in closely with the conceptual approach of the NSDI and of state-level Framework efforts. In that regard *The National Map* and NSDI must maintain a strict focus on those data sets that are truly needed for Framework. There is a tendency for Framework efforts to become somewhat side-tracked into including datasets that are easy, politically popular at the moment, or otherwise "of the moment." While the current focus on homeland security is laudable, necessary, and important and will provide some impetus for data collection and maintenance, our long-term focus on Framework data should not be lost. Nor should it be side-tracked into only security-related data.

- A key to making *The National Map* work will be ensuring that data stewardship is established at the lowest practical level of government.

Robert Marx
U.S. Bureau of the Census

- Only with a willingness on the part of every organization to "give a little" can the nation achieve this highly desirable objective. This project is large enough that if each organization adopts the "give a little" mentality, the benefits, direct and indirect, to it and the nation will be enormous.

- The concepts documented in the report are worthy and attainable (with the possible exception of the goal for seven-day vintage on a national scale). Others have raised the issue of "currentness" being the source of "value" to some potential data suppliers. In the context of adopting the "give a little" mentality, this is an issue on which compromise clearly is needed. Federal agencies, and I suspect most state/local/tribal agencies governed by open records statutes, generally cannot abide requests for licensure/copyright/royalties.

- The concept of "update only" databases likely is technically possible, but will require much work on the part of data producers, data users, and software vendors. Implementation of the concept requires rigid adherence to some sort of "time stamping" mechanism on the part of every data provider, similar time stamping on the part of every data user (to know when they last acquired updates from a data provider), and software that uses both items of information to determine which specific changes qualify for transfer. Simple in concept, probably not simple in execution.

Anne Hale Miglarese
National Oceanic and Atmospheric Administration (National Ocean Service)

- The National Ocean Service strongly supports *The National Map* vision and is eager to see this initiative succeed.
- Several ongoing federal initiatives are working toward similar goals [as *The National Map*], such as the Geospatial One-Stop initiative. It is not clear how *The National Map* will coordinate with this effort.
- The fact that *The National Map* has an almost entirely terrestrial focus raises concerns for the National Ocean Service. Because U.S. resource jurisdiction extends to the 200-nautical-mile exclusive economic zone, a *National Map* should include data that describe ocean resources and boundaries.
- Has the USGS given consideration to the fact that *The National Map* may significantly alter and potentially expand their customer user base? Redefining the primary customer could help to identify which data products to include in *The National Map*.
- The next step in moving forward with *The National Map* vision would be to develop an implementation strategy or business plan to help clarify some of the issues described in subsequent sections, as well as outline areas for partnership participation. It would also be helpful to prioritize tasks and provide a timeline for each component in order to make the transition from vision to reality.
- While it is possible to merge data of various sources and types to a common form, the term "high resolution" should be defined as well as

the definition of the resolution of the data for various types (urban, rural, hydrology).

▪ If the system is to be distributed with updates provided from remote servers, how will the USGS mandate and enforce data quality and accuracy? A detailed maintenance plan should be developed that creates an established protocol for data updates, verification, and maintenance and assigns responsibility for the completion of each task.

▪ Developing a national land-cover classification scheme that is useful to a wide customer base will be difficult.

Scott Oppman
Oakland County, Michigan

▪ A successful *National Map* proposal must include local governments in its primary mission, not as a secondary partner. The following tenets, or principles, of local government further support this premise:

1. Local governments manage the most accurate and current source of information, whether it's automated or in a hardcopy format.

2. Local governments are always the first responder to an emergency, natural disaster, or public inquiry.

3. Local governments have an in-depth knowledge of the customer, constituents, and local politics. In addition, customers or constituents generally rely on local governments to resolve specific issues affecting them.

John Palatiello
Management Association for Private Photogrammetric Surveyors

▪ There is an extraordinary untapped private sector that can contribute to *The National Map* and help the USGS [achieve] its national mapping goals.

James Plasker
American Society of Photogrammetry and Remote Sensing

▪ This is a massive undertaking! The very core concept of *The National Map* requires the full participation of tens of thousands of individuals and organizations, their managers and employees, and even members of the public as a whole, in a carefully choreographed effort that will maximize access to the best and most recent data while providing an adequate, if intangible, return to the collaborators so as to retain a sufficient level of interest and commitment to ensure continuing data flow. This is likely to be much more of a social engineering challenge than a scientific or technical challenge, and will require a complete overhaul of the current USGS executive and managerial mindset.

- USGS prides itself on being a science agency; their motto is "science for a changing world." Yet there is very little traditional "science" in *The National Map* concept in the way most USGS executives and managers understand earth science.

- Without a changed mindset from top to bottom at USGS it will not be possible to obtain or sustain the necessary fiscal, technical, human, and political resources that will be critical to achievement of this vision. One only must ask the question "How will *The National Map* be prioritized within internal USGS funding requests?" to understand the most elemental of challenges.

- If there is no clear evidence that participation in *The National Map* will yield immediate or short-term program benefits to that organization, then the chances of partnership participation dwindle significantly.

- For the concept to work, every organization with relevant data must think first of the greater good to be derived from *The National Map*, rather than what their small piece of that map is worth to them.

- Clearly the benefits of *The National Map* far outweigh the investment necessary to accomplish the concept. Clearly the vision is stimulating and exciting, and probably even doable. However, without adequate investment in this endeavor, the nation will not see those benefits.

J. Milo Robinson
Federal Geographic Data Committee (FGDC)

- The key to realizing this simple but powerful vision is coordination. Unfortunately geographic data coordination is a difficult problem. It is a problem that has long been recognized and dates back to the 1840s.

- The 2002 revision of Circular A-16 contains many changes from the previous 1990 version. Some of the key changes are that the circular expands responsibilities to include more government programs, not just the traditional mapping programs . . . . Perhaps more importantly, a new section incorporating the National Spatial Data Infrastructure (NSDI) has been added.

- Circular A-16, as well as activities surrounding e-government provides policy guidance to *The National Map*.
- The characteristics of spatial data need to be so well known that they can be run independently by different organizations, yet yield the same basic information in an easily understood manner both to people and computers. *The National Map* must drive the development of geospatial primitives like clocks help us to tell time.
- Web mapping services (WMS) offer great potential to simplify the organization of spatial information in a meaningful way . . . . FGDC is working closely with the Open GIS Consortium to implement WMS in the Geospatial One-Stop initiative. *The National Map* has implemented some WMS with success. This approach should be accelerated.
- Circular A-16 identifies two themes, geologic and biologic, that are not specifically included in *The National Map*. This could be an oversight in promoting integrated science within USGS.
- The USGS needs to provide national leadership. The United States lacks a strong mapping authority. This is evident by the 1998 NAPA recommendations for consolidation of basic mapping functions into a single agency. Leadership can stimulate coordination between the disparate efforts of multiple agencies resulting in a virtual consolidation of agencies and a strong multipurpose spatial data infrastructure that is needed for *The National Map* as well as the NSDI.
- *The National Map* calls for a federal advisory committee—a new committee. Perhaps USGS efforts in *The National Map* could make use of FGDC. Should specific limitations of a current FGDC be problematic, then USGS should work to change them.
- Recently OMB has urged the creation of joint business plans using OMB Circular 11 Exhibit 300 submissions . . . . A joint business plan involving multiple federal agencies, that complements the e-government initiative Geospatial One-Stop as well as individual agency mission needs, would clearly be well received by OMB. This is a new opportunity to advance *The National Map* and NSDI
- In addition to the *National Map* state liaisons, USGS should look to develop a network of State data advisors in collaboration with FGDC agencies. Several FGDC agencies have state presence ... NGS, NASA, USDA, and BLM.
- *The National Map* will need to develop certification procedures, processes, and roles that foster a coordinated spatial data infrastructure. It is possible that state spatial data advisors could work with state and

local partners, as well as volunteers, so they become certified providers of *National Map* data or services.

▪ Enterprise architecture should play an important role in the development of *The National Map* as well as the NSDI. Currently OMB is working to develop reference models for federal enterprise architecture (FEA). *The National Map* is predicated upon the FEA.

▪ It is time to transform USGS's National Mapping "from a mapping service organization to the federal agency responsible for structuring and coordinating the geographic or spatial component of the national infrastructure." (From Spatial Data Needs: The Future of the National Mapping Program [NRC, 1990]).

Curt Sumner
American Congress on Surveying and Mapping, Inc. (ACSM)

▪ ACSM contends that any data collected should include metadata acknowledging how the data is collected and to what standard of accuracy. This is a critical element for any body of data so that users will know what level of confidence to place in it.

▪ The most common datum utilized is geographic coordinates (latitude and longitude) that can be easily converted to state plane coordinate system values. The National Spatial Reference System provides a network of highly accurate horizontal and vertical reference points on which positions should be based. Standards for data collection also help to eliminate duplication and allow for multiple use of geographic data. ACSM would very much like to assist USGS in developing these standards for *The National Map.*

▪ Partnerships with the public are discussed on page 15 of the [vision] document. This section proposes the use of a volunteer force predicated on the "anticipated widespread availability of Global Positioning System capabilities in personal devices," and a training program for that force. ACSM feels that further discussion of this concept is warranted with regard to the anticipated accuracy of data collected and the potential end use of the data. Liabilities associated with unintended, and even unauthorized, collection and use of data such as that to be depicted on *The National Map* need to be considered. In fact, some states have developed rules that outline what types of GPS data collection must be conducted by licensed professional surveyors because of incorrect data provided by well-meaning individuals.

Eugene Trobia
National States Geographic Information Council

- Build partnerships to utilize states as area integrators (concept from the NSDI Framework).
- Update *The National Map* based on transactions, not snapshots.

Ian Von Essen
Washington State Geographic Information Council

- The events of September 11th only underscore that all of us have a real need for a consistent, accurate, and reliable set of geographic data that is readily available and accessible to all levels of government. It is imperative that our first responders have a single consistent mapping framework, ensuring rapid deployment in emergency situations. We need to be proactive in minimizing the confusion that our first responders often face with the current state of incompatible maps and data. For us to serve the needs of the twenty-first century, we need a USGS-supported *National Map*, just as the USGS's topographic maps aptly served the needs of the past century.

John Voycik
Greenhorn and O'Mara, Inc.

Digital data sources are extremely useful for many applications. There is now and will be in the foreseeable future an ongoing reliance on the paper (lithographic) USGS quadrangle map sheets.

Appendix E

Acronyms and Initialisms

ASCS   Agricultural Stabilization and Conservation Service
BLM   Bureau of Land Management
C-CAP   Coastal Change and Analysis Program
DEM   Digital Elevation Model
DLG   Digital Line Graphs
DOI   Department of the Interior
DOQ   Digital Orthophoto Quadrangle
DOT   Department of Transportation
DRG   Digital Raster Graphics
EPA   Environmental Protection Agency
ESRI   Environmental Systems Research Institute
FDLP   Federal Depository Library Program
FEMA   Federal Emergency Management Agency
FGDC   Federal Geographic Data Committee
GAO   General Accounting Office
GCDB   Geographic Coordinate Database
GIS   Geographic Information System
GNIS   Geographic Names Information System
GPO   Government Printing Office
GPS   Global Positioning System
ISO   International Organization for Standardization
MAF   Master Address File

MSC   Mapping Science Committee
NAD 83   North American Datum of 1983
NAPA   National Academy of Public Administration
NAS   National Academy of Sciences
NAVD   88 North American Vertical Datum of 1988
NCGIA   National Center for Geographic Information and Analysis
NDCDB   National Digital Cartographic Data Base
NDOP   National Digital Orthophoto Program
NILS   National Integrated Land System
NIMA   National Imagery and Mapping Agency
NMD   National Mapping Division
NMP   National Mapping Program
NOAA   National Oceanic and Atmospheric Administration
NRC   National Research Council
NRCan   Natural Resources Canada
NRCS   Natural Resources Conservation Service
NSDI   National Spatial Data Infrastructure
NSGIC   National States Geographic Information Council
OGC   Open GIS Consortium
OMB   Office of Management and Budget
TIGER   Topologically Integrated Geographic Encoding and
           Referencing
USDA   United States Department of Agriculture
USGS   United States Geological Survey